SpringerBriefs in Environment, Security, Development and Peace

Volume 29

Series Editor

Hans Günter Brauch, Sicherheitspolitik, Peace Research & European Security Studies, Mosbach, Baden-Württemberg, Germany

More information about this series at http://www.springer.com/series/10357
http://www.afes-press-books.de/html/SpringerBriefs_ESDP.htm
http://www.afes-press-books.de/html_ESDP29.htm

Sajal Roy

Climate Change Impacts on Gender Relations in Bangladesh

Socio-environmental Struggle of the Shora Forest Community in the Sundarbans Mangrove Forest

Sajal Roy
Institute for Culture and Society
Western Sydney University
Penrith, Sydney, NSW, Australia

Department of Women and Gender Studies
Begum Rokeya University, Rangpur
Rangpur, Bangladesh

More on this book is at: http://www.afes-press-books.de/html_ESDP29.htm

ISSN 2193-3162   ISSN 2193-3170 (electronic)
SpringerBriefs in Environment, Security, Development and Peace
ISBN 978-981-13-6775-5   ISBN 978-981-13-6776-2 (eBook)
https://doi.org/10.1007/978-981-13-6776-2

Library of Congress Control Number: 2019932620

The copyright of this photo belongs to Sajal Roy.

Editor: PD Dr. Hans Günter Brauch, AFES-PRESS e.V., Mosbach, Germany

English Language Editor: Dr. Vanessa Greatorex, England

This Springer imprint is published by the registered company Springer Nature Singapore Pte Ltd.
The registered company address is: 152 Beach Road, #21-01/04 Gateway East, Singapore 189721, Singapore

Preface

The Wrath of this Tempest: Disasters, Human Security and Gendered Relations

The ecologies of our lifeworlds are not simply externalities or background features of social life, as many modern accounts would have it. Rather, lived ecologies are part of the grounding condition of our social being. We are all embedded in changing ecologies. For parallel reasons, human security should not be treated as an add-on to core security concerns such as military and state security. Human security is so much more than a few 'complementary' factors such as food security, economic security or personal security, as the *modern* notion of 'freedom from want' would describe it (Cameron 2014). Rather, bringing ecologies and security together, embedded human security should be the basis of thinking about all security, including disaster management.

In this book on the indigenous people who live with the Sundarbans mangrove forest of coastal Bangladesh, the words of the people cry out for such an integrated understanding of both ecology and human security. Theirs are the words repeated by many indigenous peoples as, across the globe, they live complex intersecting lifeworlds. Customary ecological embeddedness meets with traditional cosmologies and modern incursions of knowledge. The forest is, in *customary* terms, 'our lives, our future and our love'. Or as another Shora person expressed it, the mangrove forest 'is like my son or daughter. Our existence is inconceivable without it'. But the forest is also, in *traditional* cosmological terms, 'the almighty God [who] would protect us from the wrath of this tempest', and, in *modern* terms, it is 'our natural oxygen factor' and source of vitamins. Expressed in modern geographical measurement, the Sundarbans forest is the worlds' largest tidal mangrove zone of 10,000 km^2, crossing the nation-states of Bangladesh and India, part of the Ganges River Delta on the Bay of Bengal. This modern 'ecological resource' is the subject of much research and international development and conservation work, with four million Bangladeshis directly dependent on it for their livelihoods. This book gives us much more than that.

In this place of the Sundarbans, different ontologies meet each other in a contradictory tangle of engagements. Rising above that complexity, it is clear, as Sajal Roy documents, that the forest—the Sundarbans—is more than just the background context for people's lives. It is the place that constitutes the Sundarbans' peoples in all their complexity. In Marisol de la Cadena's words, the Sundarbans is another kind of *earth-being*: a named entity called upon and variably made present to us in the inter-relationality of social life.

As a logical extension of these two points—first, that the forest is not simply a background context to this story, and, second, that human security is an all-embracing consideration—the third point becomes clear. Disasters cannot be treated as simply the outcome of extreme environmental events. Disasters are not natural; they are social. The cyclones of the Bay of Bengal are no exception. Cyclones Sidr and Aila do not have the same ontological presence in this book as the Sundarbans, but at the level of the customary they are also *earth-beings*. Their contradictory presence of fear and wonder is constantly felt as something more than just the proximate source of disasters measured on the Beaufort scale. In other words, a disaster is not simply a natural event or the inevitable outcome of a force of nature. Disasters should rather be understood as the outcome of social pressures: too many people living in a region which makes livelihoods precarious; lack of support for adaptation to changing climatic conditions; poor infrastructure in managing the relationship of social/natural life, etc. All these multiple causes *in conjunction with* a natural event affecting extreme impact mean that a cyclone can be called a 'disaster'. Turning our heads to look behind us, we can say that when extreme environmental events occur in places either where humans do *not* live or where the population and their infrastructure are well prepared, the events are *not* usually called 'disasters'. And humans, tigers, crabs, trees and cyclones certainly live together in this place of intensifying disasters.

Climate change—another key theme of this book—hides this third point. As conditions of climate change act to intensify the impact of natural events such as cyclones and droughts, we all begin to think of it as nature striking back at us. And the people of the Sundarbans are no exception. When the Shora villagers invoke the Sundarbans as their protector from the 'wrath of this tempest', they are using the same language as religious-inflected modern poets, though without the individualistic and heroic overtones of romantic modernism: 'Alone in those wild storms where hardest deeds are done' (Percival 1823). Climate change now makes us all vulnerable, and no people more so than the poor (particularly women living with precarious livelihoods). This is intensified for those who have become marginalized in a market-framed society. Here, both women and men are increasingly exposed to natural events as never before. Even the greater connection that men have to the capitalist market does not help them. By focusing on Bangladesh, identified as one of the most climate change-affected countries in the world (Kreft et al. 2017), and in framing the subject of this book as gender relations in an indigenous community, Sajal Roy is telling the story of the most vulnerable people in one of the most vulnerable places in the world. The Sundarbans is a place of changing integration with nature alongside modern exploitation of place, corruption and rapacious marketeers. It is an important story.

Penrith, Sydney, NSW, Australia   Prof. Paul James
December 2018   Director, Institute for Culture and Society
                                                              Western Sydney University

References

Cameron, R. (2014), "A More 'Human' Human Security: The Importance of Existential Security in Resilient Societies", in C. Hobson, P. Bacon and R. Cameron (Eds.), *Human Security and Natural Disasters*, London, Routledge, pp. 158–180.

de la Cadena, M. (2015), *Earth Beings: Ecologies of Practice Across Andean Worlds*, Durham, NC: Duke University Press.

Kreft, S., Eckstein, D., and Melchior, I. (2017), *Global Climate Risk Index, 2017*, Bonn, German Watch.

Percival, J. (1823), *Poems*, New York, Charles Wiley.

Acknowledgements

This book originated as a revised version of my M.Phil. thesis submitted at the University of Bergen, Norway. The study on which it is based was conducted with the financial assistance of the University of Bergen. Ethnographic fieldwork was conducted in the village of Shora in south-west Bangladesh, adjacent to the Sundarbans forest. I am grateful to the female and male inhabitants of Shora, who provided me with their invaluable time and shared their stories of the Sundarbans with great enthusiasm. I very much appreciate the way they welcomed me into their homes and lives during the fieldwork. The time I passed at Shora was a highly enriching experience, which offered me a great deal of memories, which I shall treasure for all time. I am grateful to the gatekeeper Monirul Islam, who gave me his unlimited company throughout the data collection process. I am particularly indebted to Dr. Maurice B. Mittlemark for his support and intellectual guidelines from the beginning of the project to the end of the write-up. His comments, guidance and suggestions led to consistent improvement of my work. I would especially like to express my appreciation of the Meltzer Research Foundation that bore the expenses of the research conferences in which I participated.

I am thankful to *Hans Günter Brauch* for his scholarly directions on the manuscript, which helped me to convert my thesis into a book. I am also indebted to Ilaria Walker, the Commissioning Editor- Social and Behavioural Sciences, Springer, for her cordial support from the beginning of this book project till the end of the preparation of the manuscript. I greatly value the time and intellectual labour of my colleagues and friends at Western Sydney University, who proofread this manuscript. Special thanks go to my wife for her continuous support and inspiration, while I was completing the manuscript. Without her love, affection and best wishes, the challenging task of completing my book would not have been possible. I am also grateful to my parents in my every part of success.

Contents

Abbreviations

BBS	Bangladesh Bureau of Statistics
CFCs	Chlorofluorocarbons
CIFOR	Centre for International Forestry Research
FAO	Food and Agricultural Organization
FGD	Focus Group Discussion
GAD	Gender and Development
GoB	Government of Bangladesh
IFAD	International Fund for Agricultural Development
IUCN	International Union for Conservation of Nature
MoEF	Ministry of Environment and Forest
n.d.	No Date
UNDP	United Nations Development Program
UNEP	United Nations Environment Program
UNESCO	United Nations Educational, Scientific and Cultural Organization
UP	Union Parishad
UV-B	Ultraviolet B
WB	World Bank
WHO	World Health Organization
WWF	Worldwide Fund for Nature

Chapter 1
Sundarbans Forest and the Gendered Context of Cyclones: Sidr and Aila

Abstract The recurrence of cyclones as a form of extreme weather events is causing the degradation of the Sundarbans mangrove forest in Bangladesh. This study aims to discover the forest society members' perceptions and behaviours about the Sundarbans, considering the before and after dimensions of the cyclones Sidr and Aila, which occurred in the coastal district of Satkhira adjacent to the Bay of Bengal. The study predominantly focuses on the forest-dependent women and men in a village called Shora. By employing the qualitative research methods of observation, in-depth interviews and focus groups, the present study critically investigates the women's and men's detailed perceptions of the forest, its resources, and how their perceptions and interactions in the use of forest resources have been affected by Sidr and Aila. In addition, the study documents the inhabitants' notions about environmental security as it relates to the Sundarbans and their region. The study shows that the inhabitants become acquainted with the forest during their childhood through storytelling narrated by the senior family members and the elders known as *Murubee*. Although the local people follow long-established practices and beliefs when visiting the Sundarbans, the study reveals that, compared to the women of Shora, the men act in a bolder manner to gain access to the more distant and denser part of the forest throughout the year in the hope of higher cash incomes. In the pre-cyclone landscape, a few ultra-poor married, widowed and divorced women would enter the closer part of the forest to earn their livelihood. In the post-cyclone landscape, women, rather than men, harvest the forest resources in a more sustainable way. Yet, due to patriarchal attitudes and conservative perceptions of women based on religion, women gain only small benefits from forest resources as their access is confined. Furthermore, people consider Sundarbans a great source of oxygen, a provider of human security, and at the same time a protector from natural disasters. The findings suggest that in the post-cyclone context, many women are challenging long-established practices and beliefs by engaging in income-generating activities inside the forest, rather than in their homes. It also confirms that cyclone survivors prefer to earn an alternative non-forest source of income in order to protect the forest from human intervention.

Keywords Sundarbans · Cyclones · Sidr · Aila · Shora · Gendered knowledge · Environmental security

1.1 Introduction: Sundarbans Forest and Cyclones

The Sundarbans forest in the south-west of Bangladesh has undergone conspicuous ecological changes for many years. Between 1988 and 2018, recurring cyclones and associated floods, prolonged summer droughts and intense cold weather during the winter, and hazardous use of forest resources by the area's inhabitants in their quest for livelihood, have created a severe imbalance between the forest ecosystem and human life.

Previous research studies (such as Basar 2009; Bhowmik/Carbal 2011; Aziz/ Paul 2015; Ahamed/Ahamed n.d.; Mozumder et al. 2018; Islam et al. 2018) in the mangrove region reveal the nature of Sundarbans' ecology, including changes in the habitat of wild animals and the social well-being of the forest users. Cyclones, including Sidr and Aila, have threatened the lives of coastal dwellers near the Sundarbans. This geographical vulnerability is compounded as the coastal sub-populations that mainly interact with forest resources are marginalized women, and men, who, with traditional Islamic beliefs, exhibit patriarchal attitudes to the rural society. The main purpose of this book is to document the activities of men and women inside Sundarbans forest and, in particular, women's behaviour towards the forest's resources. This study focuses on the Shora community who live near the forest and reveals the importance of this community for gender study in climate change.

The extreme and contrasting weather events of cyclones and floods followed by drought and water scarcity continually threaten the lives of impoverished Bangladeshi, with the inhabitants of the south-west coastal regions of the country most affected. Research studies (such as Basar 2009; Bhowmik/Carbal 2011; Aziz/ Paul 2015; Ahamed/Ahamed n.d.; Mozumder et al. 2018; Islam et al. 2018) have already explored the suffering, vulnerability and environmental degradation of the disaster-affected regions. Between 2007 and 2009, the inhabitants of the rural Satkhira district, who are dependent on the Sundarbans forest, experienced Cyclones Sidr and Alia, both of which resulted in loss of life, injury, and damage to local infrastructures (roads, houses), fishing ponds and cultivable lands. Sidr ruined a substantial amount of trees and killed many wild animals of the Sundarbans forest, whereas Aila destroyed crops in the field, resulting in the forested area inhabitants becoming homeless and thus marginalized (Roy 2011). Due to the shortage of food, drinking water and shelter, the Government of Bangladesh, international donor agencies, as well as several national-level NGOs, provided humanitarian relief, which contributed to emergency livelihood support for the cyclone survivors in the rural villages of Satkhira[1] (Roy 2011). Due to difficulties experienced in the post-cyclone landscape, men and women in the remote villages, cut-off from the district town, felt compelled to base their livelihoods solely on the Sundarbans forest. Because of the local and global climate change and its geographical location, Bangladesh is vulnerable to natural disasters that create high levels of damage to the population, infrastructure and several sectors, including agriculture, livelihood and livestock (Solayman 2017).

Category 4 Cyclone Sidr hit south-west Bangladesh on 15 November 2007 (Paul 2009). The cyclone moved across the west coast, towards the middle of the country, with winds reaching 248 kph (155 mph), and creating tidal surges of up to 6 m (20 feet). The death toll reached 3,406 people, with another 871 declared missing and 55,000 physically injured (GoB 2008). About 1.87 million livestock and poultry perished whilst the cyclone completely or partially damaged 2.4 million acres of crops and caused a 36-hour blackout due to power outages (Natural Hazards Centre 2008). The estimated damage bill was US$1.7 billion – approximately three per cent of the Bangladesh Gross National Product (GNP) (GoB 2008). The districts of Bagerhat, Barguna, Patuakhali, Satkhira and Pirojpur were the worst impacted by the cyclone, and the private sector suffered the most damage and loss. More than two-thirds of the loss was physical and one-third was economic (Nadiruzzaman/ Wrathall 2015). According to GoB (2008), all the affected coastal districts were already experiencing high levels of poverty, and when interviewed by the emergency relief agencies, the survivors had nothing but the clothes on their backs. Intruding saltwater contaminated the drinking water in ponds, and crops (Nadiruzzaman/Wrathall 2015).

On 25 May 2009, Category 1 Tropical Cyclone Aila hit the south-western coast of Bangladesh, causing damage to the region still recovering from Cyclone Sidr 18 months earlier. 9.3 million people were affected, either losing their lives, experiencing physical injury or losing their homes, with 190 people dying and over 7,000 being injured in Khulna and Satkhira districts alone. Cyclone Sidr caused damage to crops and trees, regional infrastructure and communication systems. The disaster resulted in widespread poverty, unemployment and hunger. A large tract of the Sundarbans forestland also was severely damaged (Alamgir n.d.). The local population affected by Aila took shelter on the embankment for about 2 years, during which they relied on external aid from the government of Bangladesh, local non-governmental organizations (NGOs), international non-governmental organizations (INGOs) and other civil society organizations (Mallick et al. 2011). According to Kamal/Hassan (2018), Cyclone Aila washed away the material possessions of the people and the social life of the community. Solayman (2017) found Sidr and Aila severely impacted the agricultural (100%) and livelihood (91%) sectors, with most impacts stemming from rising sea levels, reduced livelihood opportunities, settlement damage, economic insecurity and migration problems.

---

[1]Source: http://en.banglapedia.org/index.php?title=Satkhira_District.

Gender plays a crucial role in the social life of disaster-affected poor forest communities (Enarson 1998). Nelson et al. (2010) suggest that, despite articulating environmental vulnerability and natural hazards along social, poverty and gender lines, development policy and disaster management in the Global South largely disregards gender. This has created a general knowledge gap about the impact of disasters on gender relations, and a specific gap in the literature documenting the situations of gendered relations of both cyclones Sidr and Aila survivors of south-west Bangladesh (Cannon 2002; Drolet et al. 2015; Solayman 2017). Habtezion (2016) argues that climate change is not gender-neutral, as it dispro-portionately affects women because they are more likely to suffer higher rates of post-disaster mortality, morbidity and poverty. The underlying factors that exac-erbate a woman's vulnerability include limited livelihood options, and less access to education and basic services. They also face discrimination in social, cultural and legal norms and practices, compounded by their lack of involvement in decision-making processes at local, national and international levels. Although women are key agents of change, development programmes and policies do not consider the specific needs and concerns of women; hence, women's unique knowledge is essential to ensure the effectiveness and sustainability of climate change adaptation (Habtezion 2016).

In 2015, the United Nations (2015a) established the 'Sustainable Development Goals' (SDGs), and Goal 5 of 17 seeks to 'achieve gender equality and empower all women and girls'. This goal considers gender equality not only as a fundamental human right but also as a necessary foundation for a peaceful, prosperous and sustainable world (UN 2015b). To achieve gender equality, sound policies and legislation supporting equal access to education, health care and employment are necessary, but the decisions need to incorporate the female voice; hence, women need full and effective participation and equal opportunities for leadership. Only then will there be a chance to eliminate violence, discrimination and harmful social and cultural practices, including child marriage, while providing girls and women with access to health care, and particularly supporting their sexual and reproductive health and rights. Reforms are also required not only to give women equal rights to economic resources, particularly through enabling technology, but also to recognize and value women's unpaid care and domestic work, as this combination of factors will fuel sustainable economies (UN 2015b). As this study focuses on the gendered context of a particular community near Sundarbans, the contextual implications of the research will add value to Goal 17 targeted by the UN.

Literature on gender and post-disaster studies (e.g. Roy/Bhaumik 2013; Biswas 2013; Bose 2016; Roy 2012; Agarwal 2009) draws inadequate attention to women's behaviour towards the Sundarbans forest and its resources, hence pre-senting a gap noted by Rajoana (2017) that examines cyclone-affected women's perceptions about the Sundarbans forest as a source of livelihood. Rajoana (2017) holds that women in the rural communities regularly exploit the forest biomass for their families' survival, which makes them vulnerable to environmental changes. The patriarchal society compounds this vulnerability through cultural discrimina-tion that limits the women's information about financial sustenance, resulting in social and environmental insecurity, poverty and poor health outcomes (Rajoana 2017).

The national governments of South-Asian countries are not considerate of the women's experiences in the cyclone-affected communities in the mangrove regions. Agarwal (2009) notes that although women in the forest regions of India and Nepal can contribute to forest-related knowledge, the authorities do not appreciate them. Garai (2016) remarks that Bangladeshi women often confine themselves to household labour and disaster-specific roles, such as caring for children and ailing people, and saving properties from obliteration. This book focuses on documenting gendered activities and perceptions of environmental security in the Sundarbans forest from the study of the Shora women and men's activities in the Sundarbans forest and women's use of the forest resources, before and after Cyclones Sidr and Aila.

The study answers the following questions:

- What are women's and men's perceptions of the Sundarbans, its resources and changes in the use of forest resources due to Cyclones Sidr and Aila?
- How do women actually use forest resources?
- What are women's and men's perceptions of environmental security, as it relates to the forest and their region?

1.2   Sundarbans and the Village of Shora

The Sundarbans[2] (Figs. 1.1 and 1.2) is the largest single area of tidal mangrove forest in the world, jointly owned by Bangladesh and India, with 60% located in the south-western districts of Satkhira, Khulna and Bagerhat in Bangladesh (HCoBS 2013). 6,000 ha are located in Bangladesh, equalling 4.13% of the country's total area and 38.12% of the forest department controlled land. Whilst the forest is part of the Ganges River delta, it also connects numerous branches of other rivers that produce muddy lands and small islands of salt-tolerant mangrove forest. The rivers and adjoining canal waterways cover an area of 1,757–1,864 km^2, and the unique nature and environmental behaviour of the Sundarbans has led to erosion along the major riverbanks and movement at the land–water interface of the Bay of Bengal (Aziz/Paul 2015).

This mangrove forest houses 334 flora species and several fauns, including 35 mammalian, 270 avian, 400 fish and 35 reptilian species. Sundarbans contains about 12.26 million m^3 of timber from Sundri (*Heritiera fomes*), Gewa (*Excoecaria agallocha*), Keora (*Sonneratia apetala*), Baen (*Avicennia officinalis*), Dhundul (*Xylocarpus granatum*) and Passur (*Xylocarpus mekongensis*). The forest also attracts international tourists due to the presence of the majestic *Royal Bengal Tiger*

---

[2]Sundarbans means 'beautiful forest' in the Bengali language.

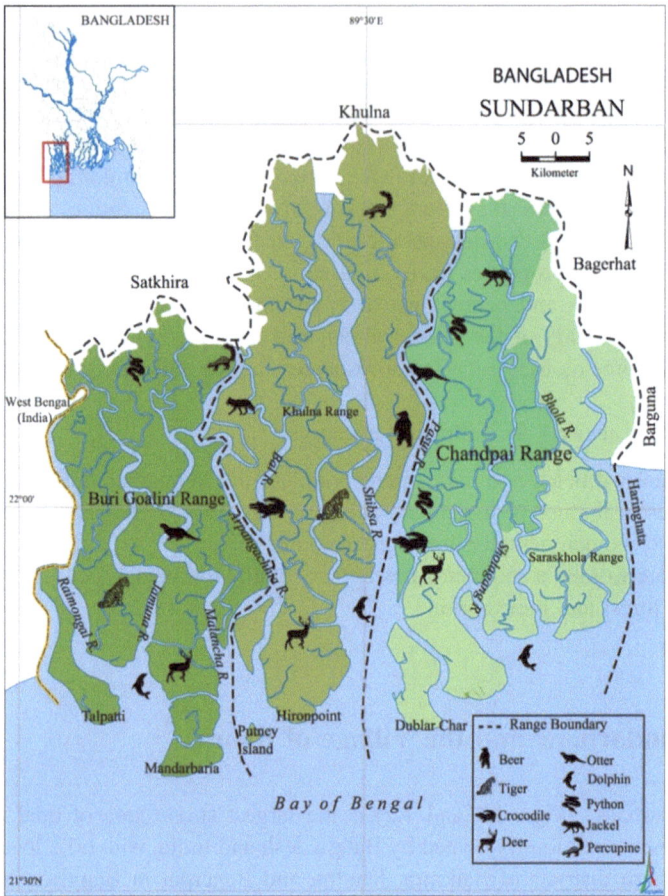

Fig. 1.1 Sundarbans, Bangladesh. *Source* www.banglapedia.org [open access]

or *Bagh*, saltwater crocodiles, several species of primates, leopards and king cobras around many small water bodies (HCoBS 2013). In 1997, the Bangladesh Sundarbans achieved UNESCO World Heritage Listing status (Dasgupta et al. 2016).

Throughout the year, the Sundarbans serves as a source of livelihood and provides ecological services to a vast majority of the people living nearby. It supplies forest products, including firewood, charcoal, fruits and honey, and fishery products, including shrimp, prawns, snails, crabs and molluscs. The calorific values of mangrove twigs contribute to making charcoal and firewood that produce heat without generating smoke (Kathiresan n.d.). The honeybees of the Sundarbans mangroves promote apicultural activities, including employment for 2,000 people in the Indian Sundarbans, who extract 111 tonnes of honey annually, accounting for 90% of annual honey production (Krishnamurthy 1990). In Bangladesh, up to 233

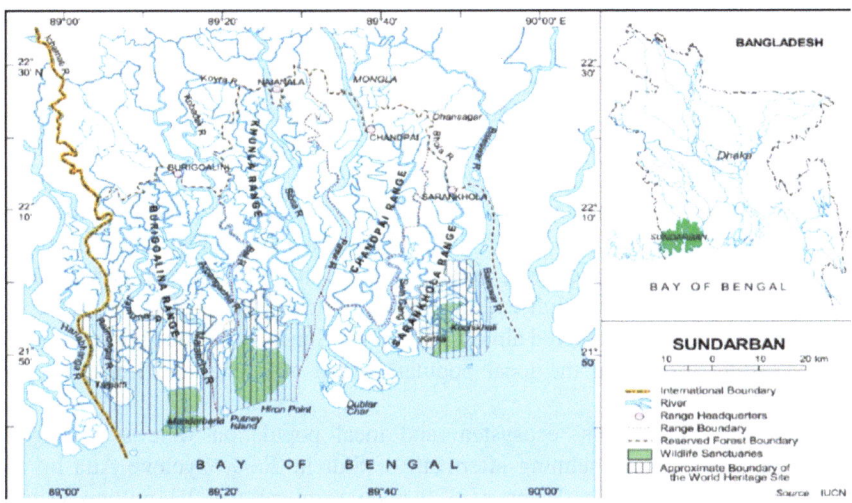

Fig. 1.2 Bangladesh. *Source* www.lged.gov.bd [open access]

tonnes of honey have been garnered annually from the western part of the mangrove forest (Aziz/Paul 2015), which is a significant increase on the 1990s, when 185 tonnes of honey and 44.4 tonnes of wax were produced (Siddiqi 1997).

In Bangladesh, the Sundarbans mangroves protect a vast area of coastal regions from UV-B radiation, the effects of global warming and the risk of cyclones, floods and coastal erosion (FD 2017). Mangrove ecosystems function as traps for sediment and sinks for absorbing pollutants such as methane, carbon dioxide and sulphur dioxide, as well as chlorofluorocarbons (CFCs). Mangrove roots hold the substrate steady and thus stabilize the ecosystem, so it can provide a source of food, breeding grounds and nurseries for fish and other fauna, support a wide variety of flora and maintain a balance between wildlife and forest animals (Kathiresan n.d.).

As the coastal dwellers near the Sundarbans rely on the livelihood[3] facilities and ecological support of the mangrove forest, they endure most environmental calamities, and at times must cope with dire situations. According to the World Bank (2018), Bangladesh is currently one of the world's most disaster-prone countries, with more than 80% of the population potentially exposed to floods, earthquakes and droughts and more than 70% to cyclones. On average, Bangladesh experiences a severe tropical cyclone every 3 years, and floodwaters inundate about 25% of the land mass, resulting in severe flooding covering 60% of the land mass (World Bank 2018).

---

[3]A livelihood comprises the capabilities, assets (including both material and social resources) and activities required for a means of living, and it is sustainable when it can cope with and recover from stresses and shocks and maintain or enhance its capabilities and assets both now and in the future, while not undermining the natural resource base (Carney 1998).

There is a consensus among climatologists that Bangladesh is most vulnerable to the impacts of climate change,[4] and because of the increase in water temperature and soil salinity between 1998 and 2008, the local population experienced decreased productivity (Basar 2009). This decrease was compounded when Cyclone Sidr struck, resulting in the deaths of 3,406 people, and a damage bill of nearly US$1.7 billion (GoB 2008 quoted in Paul 2009). The impacts are considerably worse when compared to Cyclone Gorky, which struck in the same region in 1991. Although Gorky resulted in 140,000 deaths, physical damage to coastal embankment and forest ecology, and disconnection of the road networks, Paul (2009) reveals that despite Sidr causing far fewer deaths than Gorky, it massacred the forest ecology. Sidr affected almost 45% of the Sundarbans, thus imposing more longer lasting threats on the local population and wild animals (Bhowmik and Carbal 2011).

While the Sundarbans' ecosystem and local population were struggling to recover from the overwhelming aftermath of Sidr in 2009, Cyclone Aila hit the same region, causing the deaths of 320 people among the 2.3 million victims (Kumar et al. 2010 quoted in Kamal 2013). Aila lasted longer than Sidr and, with non-stop rainfall and tidal surges; the waves lashed the embankment and submerged many villages in 15 coastal districts of Bangladesh. The tidal surges also washed away houses, crops, and livestock and livelihood sources in the affected regions.

Cyclone Aila washed away the village of Dumuria, adjacent to Shora in Shyamnagar, south of the Satkhira district. The floodwaters breached three embankments and forced villagers to seek shelter on the rooftops of mosques and primary schools (Kamal 2013). Women and children of the region were the worst affected and looked for temporary shelter in Shyamnagar Upazilla (WHO 2009). Post Aila, there is a significant change in the local population's source of livelihood and security as the inhabitants of the Shora and Dumuria villages are in a transitional phase in the recovery process of returning to their normal lives after the environmental shocks.

This study was conducted in Shora, part of Gabura Union,[5] administered by the Shyamnagar Upazilla (Fig. 1.3) approximately 45 km away from the district town of Satkhira.[6] A 75-min bus journey is required to reach Shyamnagar from the district town. Due to the muddy road network and unavailability of bus services from Shyamnagar, a motorbike or three-wheel 'Easy Bikes' are required to reach Nildumur market, a gathering place for the local people. Thereafter, a 20-min crossing of the river Kholpatura by motorboat provides easy access to the Shora at the Gabura Union, which is the closest populated area to the Sundarbans forest and

---

[4]The 4th Assessment Report (2007) of The International Panel on Climate Change (IPCC) defines the following as the main climate change impacts in the region: increased frequency of droughts and floods affecting local production negatively; sea-level rise exposing coasts to increasing risks, including coastal erosion and increasing human-induced pressures on coastal areas; and glacier melt in the Himalayas, increasing flooding and rock avalanches.

[5]Union is the lowest administrative unit of Bangladesh.

[6]Source: http://www.dcsatkhira.gov.bd/.

Fig. 1.3 Shyamnagar
Upazilla. *Source* Google
Image [open access]

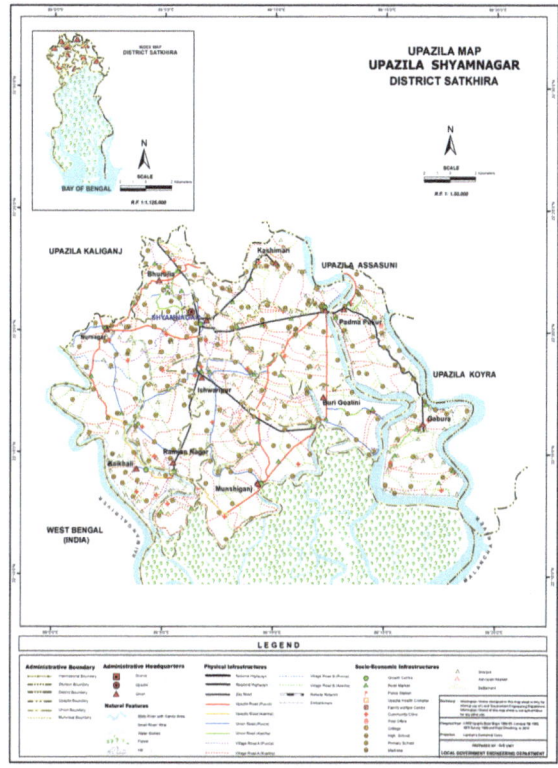

the Bay of Bengal. The entire population of the village is 5,593, of which 2,846 are women, and approximately 99% of inhabitants are Muslims (Bangladesh Bureau of Statistics 2011). The local newspaper, *Dristypat*, reports that the inhabitants of Shora struggle to earn their livelihood from the forest.

1.3 Significance of This Study

Some studies focus on women's access to, and control over, community forest resources, whilst other research concentrates on disaster victims' coping strategies, in both the northern and southern districts in Bangladesh. However, there are very few studies conducted in Satkhira and none in the village of Shora, hence, the focus of this study. At a practitioner level in Bangladesh, international development agencies have initiated and funded a large number of research projects dealing with women's rights to the forest and environmental management. Those professional research projects fail to present a comprehensive juncture between the disaster victims' knowledge of the forest and human Sundarbans relations. Consequently, this study will present a clear picture of the long-established practices and beliefs connected with the Sundarbans forest, as well as contribute to the environmental security of the forest. The dissemination of these findings will offer a new perspective in the field of customs of mangrove-forested community members and human security anticipated by the Sundarbans forest.

As per the UN declared SDG Goal 13 for Climate Action where urgent action is required to combat climate change and its impacts, this study has been conducted on the climate change effects on the Shora forest community to help achieve the targets of Goal 13 at a grass-root level. The study also incorporates women's and men's perceptions of environmental security as this will affect the protection of the Sundarbans. Considering these perceptions is necessary, as climate change is a real and undeniable threat to our entire civilization. The effects are already visible and will be catastrophic unless we act now. To protect the planet, necessary changes through education, innovation and adherence to our climate commitments must be made (UN 2015a, b). The targets include strengthening resilience and adaptive capacity to climate-related disasters, integrating climate change measures into policies and planning, building knowledge and capacity to meet climate change, implementing the UN framework convention on climate change and promoting mechanisms to raise capacity for planning and management (Globalgoals.org n.d.). The UN has made significant progress in 2017 in reaching Goal 13, such as submission of first iteration plans by seven countries and commitments from developed countries to help fund climate-related projects (Economic and Social Council 2017).

The researcher is interested in examining forest communities' standpoints, and the grass-roots activism of women and men in the Sundarbans forest area of Bangladesh. The patriarchal social structure of rural settings and the political marginalization of the mangrove regions force women to conceal their in-depth experience of cyclones and floods over the years. Due to uneven power relations, gender inequality in terms of access to forest resources, and the traditional conservative outlook in the rural villages in Satkhira, marginal women's and men's knowledge are systemically ignored. Therefore, the central focus of the study is to document the perceptions and behaviours of the cyclone-affected women and men reliant upon the Sundarbans forest.

1.4 Conceptual Clarifications of Key Terms

1.4.1 Sundarbans Mangrove Forest

Mangroves are ecologically important coastal wetland systems, and in the tropics they are especially rich in flora and fauna (Monoharan/Karuppasamy 2011). Anisur (2001: 101 quoted in Basar 2009: 6) states that:

> The importance of Sundarbans in Bangladesh's economy and regional ecosystem is enormous. More than four million people who live around this region survive on their extracting resources of this forest. Fifty thousand people from the local area rely on the forest for their livelihoods.

Sundarbans mangrove forest is the intersection of a complex network of tidal waterways and presents an excellent example of ongoing ecological processes. The area is full of a wide range of fauna, embracing 693 species of wildlife, including 49 mammals, 59 reptiles, 8 amphibians, 210 white fishes, 24 shrimps, 14 crabs and 43 mollusc species. One of the greatest attractions of this site is its varied birdlife, including 315 species of waterfowl, raptors and forest birds (e.g. kingfishers and the magnificent white-bellied sea eagle). The forest supports exceptional biodiversity in its terrestrial, aquatic and marine habitats, and is the only remaining habitat in the lower Bengal Basin for a wide variety of faunal species, including the only mangrove habitat in the word for *Panthera Tigris* (tiger). Situated in a unique bioclimatic zone within the coastal region of the Bay of Bengal, it is home to 334 plant species belonging to 245 genera and 75 families, 165 algae and 13 orchid species and is a landmark of ancient heritage of mythological and historical events. Bestowed with magnificent scenic beauty and natural resources, its high biodiversity of mangrove flora and fauna both on land and water has resulted in international recognition (UNESCO 1997).

1.4.2   Environmental Security

In 1987, the World Commission on Environment and Development (WCED) report, *Our Common Future*, used the term 'sustainable development' and introduced 'environmental security', which was tabled in the United Nations Conference on Environment and Development (UNCED) in 1992. Graeger (quoted in Freeman 2004: 9) outlines four reasons for making a theoretical and operational linkage between security and environment:

> First, environmental degradation is in itself a severe threat to human security. Second, environmental degradation or change can be both cause and consequence of violent conflict. Third, predictability and control are essential elements of military security considerations, and these are also important elements in the safeguarding of the environment. Fourth, a cognitive linkage between the environment and security has been established. It has become legitimate for mainstream politicians to speak out in favour of an environmentally responsible security policy.

1.5   Research Method

The study utilized an ethnographic method together with a phenomenological approach to answer the study questions. During five months (May 2012 to September 2012) of fieldwork, data were collected using direct observation, in-depth interviews and focus groups.

1.5.1 Data Collection

The researcher used direct observation and placed himself in a position to learn about the activities of the Shora informants by witnessing and participating in day-to-day routine activities in the forest and households of the informants. This allowed the researcher to establish rapport with the local population and immerse himself in the social and cultural practices of gleaning forest resources from the Sundarbans.

As a participant observer on a voluntary basis, involvement in the day-to-day activities with the people being studied enabled the researcher to capture not only the location of the forest and informants' wide range of salient tasks but also their patterns of interactions with the forest resources. During fieldwork, the researcher discovered insights into community members' behaviour and attitudes towards forest resources in the everyday context in a natural setting. When the researcher crossed the river, Kholpatura, to enter the field, and returned to the rented house by motorboat, he observed the condition of the forest. This allowed indirect observation and a chance to take photos. The researcher employed unstructured interviews to capture the outlooks, conducts and insights of forest-goers and non-forest users, in addition to conducting focus groups of four to a dozen participants with a view to obtaining their collective views.

The researcher began conversations in Bengali with the participants by introducing himself and explaining the objectives of the study. In order to overcome the taboo and cross-check their understandings of the forest, the researcher preferred participants who were willing to take part in the study process. During home visits, unstructured interviewing with the participants helped the researcher to discover deep insights into the ways they adapt their lives in light of Sundarbans forest.

1.5.2 Recruitment

The local inhabitants speak their own regional dialect of Bengali, which posed a challenge when trying to understand the community's traits and attitudes, especially those of women. As most of the women who use the Sundarbans forest have a low level of schooling, they adopt a conservative outlook based on the Islamic faith that influenced their upbringing. In consideration of this issue, a well-known woman was required to help communicate with these women.

1.5.3 Sampling Strategy

The main aim was to document people's activities in the mangrove forest, and women's interaction with forest resources. A total of twenty women and fifteen men constituted the purposive sample size for study. In order to document the men's perception explicitly, the research design also prioritized their concerns. This created the opportunity to obtain a detailed answer to the first study question.

1.5.4   Types of Gathered Data

1.5.4.1   Observation

For about 4 months (1 May 2012 to 31 August 2018), from dawn until 9 pm, the researcher spent time with the community members to understand their attitudes, behaviour and daily lives with the forest. The local population accepted the researcher easily, possibly due to him wearing the clothing common at Shora – 'lungi' (traditional cloth for rural men) and a towel. The researcher documented the waking times of women and men, their respective household duties and their preparatory duties before going to the forest, including how they dressed. After about 3 weeks, the researcher went to the forest with the participants for non-participant observation of how they gathered and transported forest resources to market. When the researcher noted the condition of areas of the forest with limited trees and plants, the local population informed him this was due to Sidr and Aila.

1.5.4.2   In-depth Interviews

The researcher conducted 25 unstructured open-ended interviews with men and women followed by informal conversations. The researcher started discussions on three thematic areas: mangrove forest, use of the forest resources and environmental security. This form of interviewing supported the researcher's observations about the participants' livelihoods in the forest and allowed the researcher to find out how sociocultural perceptions of the Sundarbans changed over the years. During the interviews, the participants explained interesting local terminologies and long-established practices and beliefs about the forest. The researcher wrote the important quotations in the local language in field notes during the conversations and used a digital audio recorder to record the participants' opinions.

1.5.4.3   Focus Groups

The researcher arranged two focus groups at Shora and asked the group members open-ended questions to generate interactive and lively discussion. The first focus group, consisting of women and men, shared their understandings of the mangrove forest. The first focus group was closer to the central part of Shora, while the second focus group was arranged at the distant part of Shora, situated close to the

Sundarbans. This group consisted of women only, as women in this location frequently use the forest because they are dependent on the resources for their livelihood. During the discussion, the participants talked about how they use the forest, the risks they face in doing so and how society treats them.

1.5.4.4   Data Analysis

Following the transcription and translation of data from Bengali to English, the researcher used descriptive coding to summarize the data. From September to November 2012, the researcher used Ryan and Barnard's (2003) *Techniques of Identifying Themes* to explore the themes emerging from the empirical data. As the task appeared too tough and time-consuming to study connected themes in the data, Table 1.1 lists 20 basic themes identified during the initial coding. After identifying themes from the data, the guidelines from Attride-Stirling's (2001) 'Thematic Networks: An Analytical Tool for Qualitative Research' helped outline the extraction of Basic, Organising, and Global themes in Table 1.2. The Basic Themes are the lowest-order premises evident in the text, the Organising Themes are the categories of basic themes grouped together to summarize more abstract principles and the Global Themes are superordinate themes encapsulating the principal metaphors in the text as a whole. This then helped map the thematic structure for three empirical chapters – Chaps. 3–5. After developing Tables 1.1 and 1.2, it was necessary to construct a 'Thematic Diagram' that outlines the detailed information of the empirical chapters.

1.5.5   The Role of the Researcher

The researcher had to consider differences in language, gender, social standing and intellect when interacting with the female participants. As the researcher uses a different Bengali dialect to that of the local population, this may have made the participants uncomfortable and unhappy during the interviews and focus group discussion. Given the stereotypical socialization of the female participants, they felt shy around the researcher, and even more so when their husbands were outside the homestead. Although several of the women were divorced, widowed or aged, a strong religious identity and strict gender roles within the family made it difficult to gain the trust of the female participants. The researcher also considered that his intellectual and strategic suitability might have affected power relations between himself and the participants, but frequent movements to the field and playing with the participants' children on a daily basis made it possible to solve these problems.

In the thematic diagram (Fig. 1.4), the central oval, 'Climate Change Impacts, Gender Relations in Bangladesh: Socio-environmental Struggles of the Shora Forest Community in the Sundarbans Mangrove Forest', shows the key theme of the thesis. The central oval connects three sub-ovals surrounded by a few

Table 1.1 From codes to themes

Codes (step 1)	Issues discussed (step 1)	Themes identified (step 2)
Entry to the forest	Local beliefs and practices	1. Sundarbans in several names
	Holy place	2. Forest-going
	Islam	3. Long-established beliefs and practices
	Short period of time	
	Long period of time	
	High water/low water	
	Extended period of time	
	Muscular power	
Widow	Tiger attack	4. Categories of women
Jele-Baoalie	Husbandless	5. Women's access to forest
Divorced women	Small branches of rivers and canals	
	Polygamy	
Male-headed households	Brotie, an NGO	6. Men's access to forest
	Denser part of the forest	
	Challenge taker and earner of livelihood	
	Males are viewed as banks	
Forest resources	**Trees**	7. Resources in the Sundarbans
	Gayoa, Bain, Dundol, Posur, Hetal, Sundori, Kawra, Goalpata and Goran	8. Local cultural practices
	Fishes	
	Chati, RenuPona, Powa, Tangra, Vetke, Passea, Vangan, Med	
	Others	
	Honey, Deer, Royal Bengal Tiger, Crocodiles, Snakes, Olives and Crabs	

(continued)

Table 1.1 (continued)

Codes (step 1)	Issues discussed (step 1)	Themes identified (step 2)
Corruption	Males pay bribes to the forest officer	9. Corruption in the forest office
	Permission	
	Females rarely receive permission	
	Forest ranger	
Deforestation	Flagged and non-flagged boats	10. Forest for future generations
	Unplanned cutting down of trees	11. Before and after dimensions of Cyclones Sidr and Aila
	Cyclones: Gorky, Sidr and Aila	
	Excessive extraction of forest resources	
Shora women and nature	Biodiversity loss	12. Shora women's interactions with the forest resources
	Closer connection	
	Sundarbans forest as family and NGO training	
Shora women as agents of the local environment	Ecological ethics	13. Women's indigenous knowledge
	Resource-saving techniques	
Gender division of labour	Unpaid domestic work	14. Active participation of Shora women
	No choice and options in familial decision-making	15. Women's invisibility in the local market
	Limited scope for applying agency	
	Limited access to credit and control over it	
Local Ecology	Food and salinity in the water	16. Main source of cash income
	Infertile land	17. Existing environmental resources
	Oxygen factory and sources of vitamins	18. Trees are silent contributors
	Increasing temperature in the summer and severe cold in the winter	19. Cyclone centre
	Impacts of Sidr and Aila	20. Tree plantation
	Resistant to cyclones	

Source The author

Table 1.2 From basic to organizing to global themes

Themes as basic themes	Organizing	Global themes
1. Sundarbans in several names	Acquaintanceship with the forest	Narratives of the Sundarbans forest at Shora
2. Forest-going		
3. Long-established beliefs and practices	Forest-going practice Forest serves basic needs of life	
4. Categories of women	Illegal activities inside Sundarbans	
5. Women's access to forest		
6. Men's access to forest		
7. Resources in the Sundarbans		
8. Local cultural practices		
9. Corruption in the forest office		
10. Forest for future generations		
11. Before and after dimensions of Cyclones Sidr and Aila		
12. Shora women's interactions with the forest resources	Primary resource user Decision-making between market and women gathers Women as natural conservators of the Sundarbans	Women's use of the Sundarbans forest resources
13. Women's indigenous knowledge		
14. Active participation of Shora women		
15. Women's invisibility in the market		
16. Main source of cash income	Human security Informants call for environmental security Women's special care of Sundarbans forest	Human security, Sundarbans and survival at Shora
17. Existing environmental resources		
18. Trees are silent contributors		
19. Cyclone centre		
20. Tree plantation		

Source The author

rectangles. The sub-oval 'Narratives of Sundarbans: Informants' Perspectives' at the top of the central oval is the title of the first empirical chapter, 'First Global Theme', that attempts to answer the first study question in detail. The items listed in the six rectangles surrounding this sub-oval are the subsections 'Organized Themes' of the chapter, which tend to elaborate the discussion with the support of the 'Issues discussed' section presented in Table 1.1. The sub-oval at the bottom right, 'Women's behaviour towards Sundarbans forest', presents the essence of the second empirical chapter, 'Second Global Theme', for answering the second study question. The items listed in the rectangles adjacent to this sub-oval are the 'Organized Themes' of the chapter. In addition, the sub-oval at the bottom left,

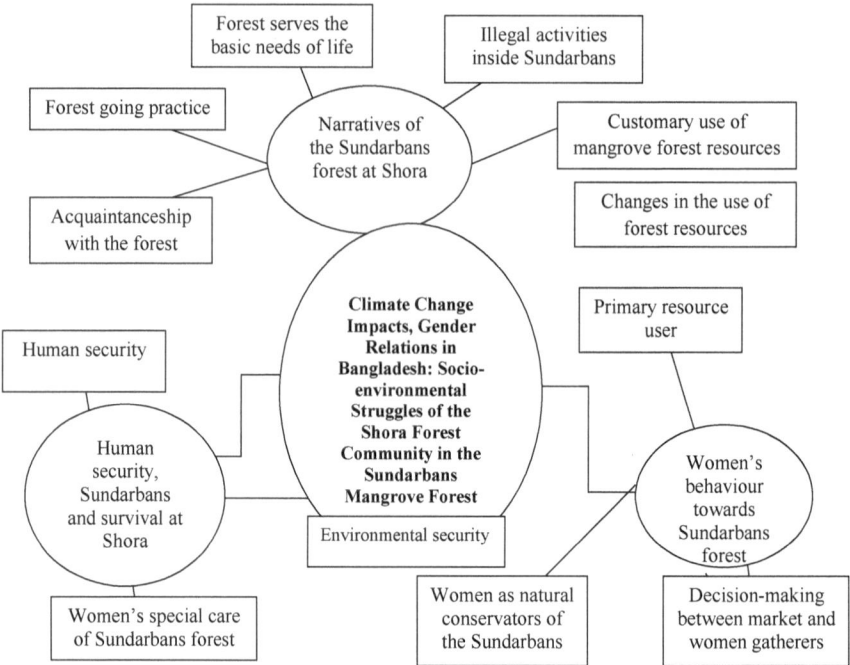

Fig. 1.4 Thematic diagram constructed based on Tables 1.1 and 1.2. *Source* The author

'Human security, Sundarbans and survival at Shora', presents the third empirical chapter, 'Third Global Theme', in which the researcher highlights the detailed answer to the third study question. The three rectangles' 'organized themes' encompassed by the sub-oval list three key points handled in subsections of the chapter to provide a more detailed discussion of the relevant issues.

1.5.6 Ethical Considerations

As the fieldwork was accomplished on behalf of a Norwegian university, the Norwegian Social Science Data Services (NSD) required clarification. After talks with the programme director of Gender and Development at the University of Bergen, the researcher did not have permission from the NSD; hence, the data was not permitted to disclose any information which might identify the participants. This made the researcher determined to sustain the ethical standard of the fieldwork.

At the beginning of the interviews, the participants were very curious about the researcher, hoping that he would provide relief and aid, including building a comfortable shelter in the region for the disaster-affected people. The participants initially wanted to be video recorded, so that the recording could be played back in

Norway to show the suffering caused by Sidr and Aila. To achieve the ethical standard, the researcher described the purpose of the study and explained that instead of filming them, he would write about their life experiences in relation to the forest and disasters, and their longing for a secure home environment. After providing this information, they consented to participate in the study.

Asking the names of people during formal or informal conversation is decorum in the rural society of Bangladesh. Shora people like their visitors to address them with honour. Consequently, during the interviews with the participants, the researcher had to ask for their names and their residential history, and introduced himself in the local language. To adhere to ethical research practice and preserve confidentiality, the research maintained security of the audio recording of individual interviews, focus group interviews, transcriptions of interviews and field notes.

1.6   Chapters in the Book

This book consists of five chapters. Chapter 2 reviews the literature on three thematic areas: (a) women's knowledge about the forest; (b) gender relations in the forest-related research; and (c) gender in the environmental security discourse. The discussion clarifies how social forces, encouraged by patriarchal attitudes, downplay women's conventional knowledge and experiences about the forest and usage of its resources. This invisibility of women continues in the domestic sphere, where they are compelled to restrain their voice in the familial decision-making process. This chapter also discusses the gendered perspective of environmental security, describing how resource scarcity and disaster cause environmental degradation. The core of the chapter uses the Standpoint Theory and the Feminist Political Ecology Theory to unpack the purpose and themes of the research. The chapter discusses the historical growth of the Standpoint Theory and presents the arguments of Harding (1993, 2004a, b, 2008) and Mohanty (1991). Harding argues that the Standpoint Theory schematically outlines the epistemological base of feminist knowledge and explores the politics against marginalized women's knowledge production at the grass-roots level. Mohanty maintains Third World women's knowledge production that helps to explain the struggles of the female participants in the study context. Furthermore, the arguments of Feminist Political Ecology Theory offer an overall understanding of female forest users' social position, their domestic struggles and how they lose the cash income gained from the forest resources.

Chapter 3 reflects the participants' detailed understanding of the Sundarbans forest. The discussion clarifies how the forest area inhabitants acquire their initial understanding of the forest and maintain the long-established practices and beliefs for visiting and using the Sundarbans. As some of the local population use the forest in legal and illegal ways, the chapter elaborates on the changing relationship with the customary use of the forest resources and interprets the findings through the Standpoint Theory, with the integration of relevant literature.

Chapter 4 describes how female participants in the study site deal with the forest and its resources, both at the Sundarbans and at home. The findings confirm that women in the study site are the primary users of the gathered forest resources, but their decision-making is situated between the marketplace and collection. Despite their vulnerable position at Shora due to the impacts of cyclones and the patriarchal attitudes of the society, the female participants are motivated to conserve the forest for their sons and daughters and seek alternative non-forest sources of livelihood.

Chapter 5 considers the forest pre- and post-Cyclones Sidr and Aila, and discusses the participants' views on human security in relation to their survival and the forest. This chapter explains that each participant holds knowledge about the benefits of human security and the threats to it, and that before Cyclones Sidr and Aila, Shora had a wealthy ecology for the local population. The chapter further explains how the forest ecology was adequately equipped to maintain a balance between wildlife and climatic threats but the participants' narratives reveal that after the cyclones, the forest-centred traditional occupations have changed, and depletion of the forest has caused threats to life.

Chapter 6 outlines the summary of each chapter and critically discusses the interpretation of the theories and implications of the current study. The last section of this chapter offers ideas for further research, including the mangrove regions of Bangladesh.

References

Agarwal, B. (2009), 'Gender and Forest Conservation: The Impact of Women's Participation in Community Forest Governance', *Ecological Economics*, Vol. 68, No. 11, pp. 2,785–2,799.

Ahamed, U.M. and Ahamed, F. (n.d.), *Control of Illegal Activities in the Sundarbans Forest of Bangladesh: Overview of the Regulatory Aspects*, Seminar Paper, Department of URP, Khulna University, Khulna, Bangladesh.

Alamgir, M., *Still Feeling the Toll of Cyclone Aila*, Islamic Relief, United Kingdom, viewed 19 June 2018. https://www.islamic-relief.org.uk/cyclone-aila-five-years/.

Attride-Stirling, J. (2001), 'Thematic Networks: An Analytic Tool for Qualitative Research', *Qualitative Research*, Vol. 1, No. 3, pp. 385–405.

Aziz, A. and Paul, A.R. (2015), 'Bangladesh Sundarbans: Present Status of the Environment and Biota', *Diversity*, Vol. 7, No. 3, pp. 242–269. https://doi.org/10.3390/d7030242

Basar, M.A. (2009), 'Climate Change, Loss of Livelihood and Absence of Sustainable Livelihood: A Case Study of Shyamnagar, Bangladesh', Master's Thesis, Asian Studies, Lund University, Sweden.

Bhowmik, A. and Carbal, P. (2011), 'Damage and Post-cyclone Regeneration Assessment of the Sundarbans Botanic Biodiversity Caused by the Cyclone Sidr', *1st World Sustainability Forum*, pp. 1–16.

Biswas, B. (2013), '"The God of Death Takes Half of Our Children": Health of Women and Children in the Sundarbans Islands', *Journal of Health Care for the Poor and Underserved*, Vol. 24, No. 2, pp. 730–740. https://doi.org/10.1353/hpu.2013.0069.

Bose, S. (2016), 'Through the Lens: Empowering Women in Vulnerable Communities to Voice their Concerns', *Stories of Change*, No. 4, Brighton: Future Health Systems. http://opendocs.ids.ac.uk/opendocs/handle/123456789/12958.

Cannon, T. (2002), 'Gender and Climate Hazards in Bangladesh', *Gender & Development*, Vol. 10, No. 2, pp. 45–50. https://doi.org/10.1080/13552070215906.

Dasgupta, S.; Huq, M.; Mustafa, M.G.; Sobban, M.I.; Wheeler, D. (2016), *Impact of Climate Change and Aquatic Salinization on Fish Habitats and Poor Communities in the Southwest Coastal Bangladesh and Bangladesh Sundarbans*, World Bank Policy Research Working Paper 7593, viewed 22 June 2018. https://openknowledge.worldbank.org/handle/10986/24135.

Drolet, J.; Dominelli, L.; Alston, M.; Ersing, R.; Mathbor, G.; Wu, H. (2015), 'Women Rebuilding Lives Post-disaster: Innovative Community Practices for Building Resilience and Promoting Sustainable Development', *Gender & Development*, Vol. 23, No. 3, pp. 433–448. https://doi.org/10.1080/13552074.2015.1096040.

Economic and Social Council (2017), *Progress Towards the Sustainable Development Goals: Report of the Secretary-General*, United Nations.

Enarson, E. (1998), 'Through Women's Eyes: A Gendered Research Agenda for Disaster Social Science', *Disasters*, Vol. 22, No. 2, pp. 157–173.

Forest Department (2017), *People's Republic of Bangladesh*, viewed 20 June 2018. https://goo.gl/77xvZJ.

Freeman, M.A. (2004), 'Environmental Security in the Global Capitalist System: A World-Systems Approach and Study of Panama', MA Thesis, Department of Political Science in the College of Sciences, University of Central Florida, Orlando, Florida, US.

Garai, J. (2016), 'Gender Specific Vulnerability in Climate Change and Possible Sustainable Livelihoods of Coastal People: a case from Bangladesh', *Journal of Integrated Coastal Zone Management*, Vol. 16, No. 1, pp. 79–88. https://doi.org/10.5894/rgci656.

Habtezion, S. (2016), *Gender, Climate Change Adaptation and Disaster Risk Reduction*, Training Module 2, UNDP.

High Commission of Bangladesh, Singapore (2013), *The Sundarbans: World's Largest Mangrove Forest Where Lives the Royal Bengal Tiger*.

Islam, M.M.; Sunny, A.R.; Hossain, M.M.; Friess, D.A. (2018), 'Drivers of Mangrove Ecosystem Service Change in the Sundarbans of Bangladesh', *Singapore Journal of Tropical Geography*, Vol. 39, No. 2, pp. 244–265. https://doi.org/10.1111/sjtg.12241.

Kamal, M.M.A. (2013), 'Livelihood Coping and Recovery from Disaster: The Case of Coastal Bangladesh', *Current Research Journal of Social Sciences*, Vol. 5, No. 1, pp. 35–44.

Kamal, M.M.A. and Hassan, S.M.M. (2018), 'The Link between Social Capital and Disaster Recovery: Evidence from Coastal Communities in Bangladesh', *Natural Hazards*, pp. 1–18, Dordrecht: Springer Netherlands. https://doi.org/10.1007/s11069-018-3367-z

Kathiresan, K. (n.d.), *Importance of Mangrove Ecosystem*, Centre for Advanced Studies in Marine Biology, Annamalai University, India.

Krishnamurthy, K. (1990), 'The Apiary of Mangroves', in: D.F. Whigham, D. Dykyjoya and S. Hejnyeds (eds.), *Wetland Ecology and Management: Case studies*, Dordrecht: Kluwer Academic Press, pp. 135–140.

Mallick, B.; Rahman, K.R.; Vogt, J. (2011), 'Coastal Livelihood and Physical Infrastructure in Bangladesh after Cyclone Aila', *Mitigation and Adaptation Strategies for Global Change*, Vol. 16, No. 6, pp. 629–648.

Mozumder, M.M.H.; Shamsuzzaman, M.; Nabi, R.U.; Rashid, A.H.A. (2018), 'Socio-Economic Characteristics and Fishing Operation Activities of the Artisanal Fishers in the Sundarbans Mangrove Forest, Bangladesh, *Turkish Journal of Fisheries and Aquatic Sciences*, Vol. 18, pp. 789–799. https://doi.org/10.4194/1303-2712-v18_6_05.

Nadiruzzaman, M. and Wrathall, D. (2015), 'Participatory Exclusion – Cyclone Sidr and its Aftermath', *Geoforum*, Vol. 64, pp. 196–204.

Nelson, V.; Meadows, K.; Cannon, T.; Morton, J.; Martin, A. (2010), 'Uncertain Predictions, Invisible Impacts, and the Need to Mainstream Gender in Climate Change Adaptations', *Gender & Development*, Vol. 10, No. 2, pp. 51–59. https://doi.org/10.1080/13552070215911

Natural Hazards Centre (2008), 'Cyclone Sidr – Bangladesh', *Natural Hazards Observer*, Vol. 32, No. 3, p. 4.

Paul, K.B. (2009), 'Why Relatively Fewer People Died: The Case of Bangladesh's Cyclone Sidr', *Natural Hazards*, Vol. 50, pp. 289–304.

Rajoana, J. (2017), 'Gender and Sustainable Rural Tourism: A Study into the Experiences and Roles of Local Women in the Sundarbans Area of Bangladesh', *International Journal of Humanities and Social Sciences*, Vol. 4, No. 4, World Academy of Science, Engineering and Technology. scholar.waset.org/1999.39/56095.

Roy, A. and Bhaumik, U. (2013), 'Participation of Women of Sundarbans in Fisheries Decision Making', *Journal of the Inland Fisheries Society of India*, Vol. 45, No. 2, pp. 23–27.

Roy, S. (2011), 'The Roles of Women Entrepreneurs in Land Conservation for Environmental Security in Bangladesh', MSS Thesis, Women and Gender Studies, University of Dhaka, Bangladesh.

Roy, S. (2012), 'Women Entrepreneurs in Conserving Land: An Analytical Study at the Sundarbans, Bangladesh', *Canadian Social Science*, Vol. 8 No. 5.

Ryan, G.W. and Bernard, H.R. (2003), 'Techniques for Identifying Themes', *Field Methods*, Vol. 15, No. 1, pp. 85–109.

Siddiqi, N.A. (1997), 'Management of Resources in the Sundarbans Mangroves of Bangladesh', *International Newsletter of Coastal Management – Intercoast Network*, Special Edition 1, pp. 22–23.

Solayman, H.M. (2017), 'Impacts of Cyclone on Livelihood: Study on a Coastal Community, *International Journal of Social Sciences*, Vol. 4, No. 4, pp. 56–64.

Spalding, M.; Kainuma M.; Collins, L. (2010), *World Atlas of Mangrove*, London: Earth Scan Press.

United Nations (2015a), *United Nations Sustainable Development: Gender Equality and Women's Empowerment*, viewed 6 June 2018. https://www.un.org/sustainabledevelopment/gender-equality/

United Nations (2015b), *Sustainable Development Goal 13*, viewed 20 June 2018. https://sustainabledevelopment.un.org/sdg13

Washbrook, E. (2007), 'Explaining the Gender Division of Labor: The Role of the Gender Wage Gap', *CMPO Working Paper: 07/174*, pp. 1–89, submitted to the University of Bristol, UK.

World Health Organization (2009), *Gender, Women and Health*, Regional Office for South-East Asia. Viewed 17 April 2019, https://www.who.int/gender/documents/overview_activities_2004-5.pdf.

World Bank (2018), *Bangladesh Disaster Risk and Climate Change Program*, viewed 20 June 2018, http://www.worldbank.org/en/country/bangladesh/brief/bangladesh-disaster-risk-climate-change-program.

Chapter 2
Theoretical Approaches: Gendered Knowledge in Forest, Ecology and Environment

Abstract This chapter deals with the theoretical approaches of gender knowledge in forest, ecology and environment. How women's knowledge of the forest can contribute and the relation of gender in forest-related research and environmental security discourse are been discussed here. Moreover, the chapter tries to correlate the research with two widely studied theories, namely, standpoint theory and feminist political ecology theory. It was found that women possess very significant knowledge which can contribute greatly to forest management, but have less opportunity to engage in the decision-making process. This chapter shows that gender has been a noteworthy issue in forest-related studies for a long time. Gender is also a crucial topic in environmental security discourse. The literature reviewed in the chapter talks about the epistemic value of the women's lives. Thereafter, women's knowledge has been discussed with reference to feminist political ecology theory. The final section of the chapter shows how the findings of this empirical study can contribute to the current practice in academia.

Keywords Women's knowledge · Gender · Shora · Environmental security · Standpoint · Feminist political ecology

2.1 Introduction

This chapter reviews the state of current knowledge within contemporary literature in three thematic areas: (a) women's knowledge of the forest, (b) gender relations in forest-related research and (c) gender in the environmental security discourse. The chapter also reflects on the fundamental insights of two theories: standpoint theory and feminist political ecology theory. First, arguments on how rural women develop their understanding about forest and its resources are presented, as well as on how women of the South Asian countries, in particular, Bangladesh, Nepal and India have visible connections with the forest. The second section of the discussion takes account of gender relations in forest-related research, and entails the relational aspects between women and men in light of forest resources. The third section

© The Author(s), under exclusive license to Springer Nature Singapore Pte Ltd. 2019
S. Roy, *Climate Change Impacts on Gender Relations in Bangladesh*,
SpringerBriefs in Environment, Security, Development and Peace 29,
https://doi.org/10.1007/978-981-13-6776-2_2

sheds lights on how resource scarcity, as well as environmental degradation, affects the human–environment relation and how forests contribute to the protection of an ecosystem in a given context. The next part of the discussion contains feminist scholars' views that position the epistemic value of the lives of marginalized groups of women. Then, the chapter deals with feminist political ecology theory and explores how the approach of this theory explains women's excavated knowledge in environmental resources and how they develop insights into environmental security. The concluding section – 'contributions of the study' – summarizes the themes and theories, and identifies the knowledge gap of the reviewed literature to be filled by the empirical chapters of this study.

2.2 South Asian Women's Knowledge About the Forest

Stiem/Krause (2016) suggest that knowledge of the forest and forest products is highly gender specific. Vázquez-García/Ortega-Ortega (2017) found that the poleo-harvesting practices in Mexico show that males are dominant in the institutions, and they do not recognize the rights of women. Studies suggest that though women possess important knowledge about forest use, they are ignored in local forest governance (Mai et al. 2011; Lewark et al. 2011; Djoudi/Brockhaus 2011; Brown/Lassoie 2010; cited in Vázquez-García/Ortega-Ortega 2017). According to Bose (2011), women's knowledge and opinions are ignored in legislation processes of forest governance. It was found in Brazilian Amazonia that women hardly play any role in decision-making on forest resources (Shanley et al. 2011). The case is almost the same in Uganda (Mukasa et al. 2012). It was found from the research of Vázquez-García/Ortega-Ortega (2017) that Mexican women depend on male relatives for information about the forest. Men attend the meetings and they inform female family members about the activities. Women have to depend on secondary sources for information. Therefore, women lack knowledge about forests. According to Stiem/Krause (2016), women are traditionally underrepresented at community meetings in Congo. They can hardly express their opinions. Participation in the clan meetings depends on land rights. From the research, it was found that women wanted to participate equally in the meetings and lessen the gender gap. Because of the higher workload women bear, they cannot participate most of the time in the forest management meetings. It was interesting that more men than women understood the reason why women could not participate equally. Moreover, there are 12 religious ideas that inspire women to accept their subordination to men. Religious leaders in churches promote these ideas using biblical references (Stiem/Krause 2016).

Gururani (2002) argues that women from less developed countries possess environmental knowledge, with a particular focus on biodiversity conservation. She further emphasizes that the essence of women's environmental knowledge is about soils, seed collection, forest species, biodiversity, pesticides, and so on. Women's knowledge about the forest resources is always overlooked and marginalized because men always interrupt the processes of the social construction and representation of women's knowledge as a sin, in Bankhali's forest in northern India (Gururani 2002). Gururani (2002) remarks that:

> Even though women know a great deal about the forest, men systematically render women's knowledge as non-knowledge and regard them as backward and foolish. Given the imprint of patriarchal hegemony, women, too, undermine their own knowledge and uphold men's scriptural knowledge as 'real knowledge,' illustrating the culturally contingent relations of power, between women and men, and between different women that shape the politics of knowledge (Gururani 2002: 321).

McDougall et al. (2013) conclude that women and the poor people are often vulnerable to changes in forest access. According to Tanny et al. (2017), females have become almost twice as vulnerable as males in northwestern Bangladesh, because of the climatic disasters. The research suggests that land possession, bathroom facility and use of borrowed money are the main determinants of gender vulnerability there. In developing countries, poor women suffer more due to the effects of climate change. As women are considered to be lower in society, they get less access than men to the resources and advantages. Women have less livelihood options than men. Most of the women depend on agriculture alone for their livelihood, which makes them more vulnerable to climate change. In Bangladesh, gender norms and restrictions impede women's access to resources, like land. Women faced shortages of food, poor health and sanitation facilities after the disasters (Tanny et al. 2017). However, most of the men opine that their knowledge is more profound than that of women, as men are the sources of women's knowledge about forest. According to them, this shortcoming of knowledge leads to women's lower participation in the forest governance committee. On the other hand, women suggest that they possess more knowledge than men about the forest. They supported their claim by demonstrating that they spend more time in the forest collecting firewood and cultivating land. To judge the claims of the two parties, an observation was made. The result showed that women had a broad knowledge about the forest resources (Stiem/Krause 2016). Therefore, socially constructed gendered norms and ideas affect the perceptions of women's knowledge and contributions to forest governance. These gendered ideas of men and women's lower level of education construct men as the hegemonic party of the society, and build barriers which prevent women participating equally in decision-making authorities for land and forest governance (Agarwal 2001; Sen 1990; Gurarascio et al. 2013).

Kabir/Hossain (2008) explicitly present the symbiotic relationship between nature and indigenous people. They suggest the integration of the indigenous knowledge of traditional forest resource users and the exclusionary perspectives of women's forest-related knowledge. According to Hossain/Islam (2017), forest entrants pay 10–15 times extra money, as a bribe when seeking permission to enter the forest to collect forest products. As a result, over-exploitation of forest resources occurs to offset the additional costs via illegal methods. As the women remain ignorant due to lack of financial and technical knowledge, this over-exploitation continues to harm the ecological balance and eventually leads to destitution for the communities.

'People's attitudes towards Social Forestry: A Case Study in Rajshai,' a study conducted in Bangladesh, claims social forestry is a source of capital for the community dwellers. Capital earned from the forest improves the aesthetic view of the area (Alam et al. 2012). For the increase in literacy and the resultant close contact with mass communication, the study also explores how farmers become aware of the role of social forestry for their own benefits. According to this study, each part of the tree is considered valuable in the indigenous community, because it provides food, fuel, furniture and, most importantly, medication.

Four million people in the coastal region of Bangladesh are directly dependent on the Sundarbans forest (Anon 2004, quoted in Iftekher/Islam 2004). In this view, Iftekher/Islam (2004: 142) assert that the mangroves play a pivotal role in offsetting the aftermath of cyclones. Mangroves also supply indispensable nutrients and habitats for fish and wild species. Their study points out how mangroves provide raw materials for paper, pencils, and the wood and furniture industries. Bosold (2012) analyses how power dynamics and relations between community members and forest officials affect the mangrove forest conservation decision. This study also demonstrates how gender plays an important role, as men and women, with their different positions at community level, make use of the mangrove forest differently. Women and men constitute their perceptions of the forest in various ways. For example, Siar (2003, cited in Bosold 2012) acknowledges that men place a higher value on the commercial species of fish thriving in offshore coral banks.

Siar (2003) observes that the activities of men in Honda Bay in the mangrove forest are usually deemed principal, while those of women are either ignored or considered peripheral. Siar (2003) remarks that the women are marginalized and their work is overlooked, which needs to be vigorously addressed in the study. Bosold (2012) further argues that a gender perspective assists the inclusion of the marginalized women's experience of the mangrove forest. He provides an overview of where the activities of men and women converge and diverge, both in the forest and within the household. With the depletion of the natural resources and the fast increase in population, women face dire scarcities of firewood (IUCN 2007). This obliges rural women of the northern districts Ghanche and Ghizer of Pakistan to walk a long distance to collect firewood. According to Islam (2011), forests are the most diverse and widespread ecosystem on Earth and forest products have served millions of people with innumerable essentials for their livelihood for centuries. His study on the Sundarbans forest explores four types of traditional occupations of the forest dependents: fishing, crab catching, honey collecting and harvesting non-timber forest products, such as Nipa Palm (*Nypa fruticans*).

The tradition of occupations in the Sundarbans forest areas has been utterly transmuted with the cyclonic hits of Alia and Sidr (Nasrin 2012). The lives of women in the villages in Bangladesh are shaped by the dual aspects; women are completely dependent on natural resources to provide the food, fuel, fodder, water, medicine and income-generating activities they need to survive. They have to carry out their familial responsibilities by managing and using such natural resources

(Nasrin 2012). Women act as natural resource managers, who make decisions on how to use environmental resources within the household. Conversely, poor women become the worst affected victims in the event of natural and man-made disasters. They face disasters like floods, cyclones, droughts, deforestation, soil and riverbank erosion, drying of wetlands, contamination, agro-chemicals and industrial waste, commercial shrimp cultivation and inappropriate land use. Nasrin (2012) further argues that the interconnection between women and the environment is less evident in the west, where the majority are not engaged directly in cultivating or gathering their food supply or the fuel and water they consume. The participatory process in the resource management benefits everybody in a given community, but the social reality and complex relationship between different actors impede the management of the community-owned forest (Buchy/Subba 2003). Their study addresses how an institutional model like the community forest restricts the integration of gender as a variable.

The rural women in the forest communities in Nepal are recognized as the key forest managers for their knowledge, skills, contribution, and dependency on the forest resources (Giri 2012). Giri (2012) further adds that the ambiguity between the forest tenure and the involvement of different categories of women and men in South Asian societies are poorly understood. 'The institutional mechanism and wider political context shaping the power relation in the gender perspective of the forest concerned occupations are still unclear' (Giri 2012: 2). Similarly, Halim (1999) agrees that men and women in the forest areas of Bangladesh are knowledgeable about social forestry, and hence able to meet their household needs. The professional aspects of women in this sector are unfortunately quite low compared to those of men. Based on her research in India and Nepal, Agarwal (2009) suggests that there is a positive correlation between women's participation in forest governance and sustainability. Women can express their thoughts and interests if they are included in the decision-making process. The executive committees of the groups with more women than men resulted in better outcomes in forest conditions. Also, there was less forest degradation.

In Nepal, groups with all-female executives showed an even better result. Women's presence contributes to improved forest protection and better management. Moreover, women's knowledge of plant species, resource collection methods and cooperation with other women also contribute to good forest governance. Women in the study informed the author that men need the cooperation of women to protect the forest as they need someone to run the house. They maintained that equal contributions by both sexes are necessary to achieve sound forest management (Agarwal 2009). Agarwal (2009) holds that there is a difference between men and women in forest products use. Rural women generally use fuelwood, fodder and non-timber resources for domestic purposes. Men occasionally collect timber to sell at market. Leisher et al. (2016) hold that the difference between women's and men's use of natural resources has minimal influence on local resource management. Engaging more women in forest and fishery decision-making bodies can bring a fruitful change in forest governance and conservation.

White/Martin (2002 cited in Sun et al. 2012) articulate that the rights of access and the use of forest resources by forest-dependent communities offer a strongly based livelihood, and better forest management as well as conservation. Sun et al. (2012) review national laws, with special attention to the individual, collective and public right of control over the forest and forest resources of South Asian countries. Their review emphasizes the way the national law interprets the position of men and women in society. Social customs build gender discrimination, giving priority to men as the breadwinners and undermining women as social quasi-outcasts. 'Forest dependent women seldom have secure title to forest lots or secure access to forest resources' (Sun et al. 2012: 2). Therefore, with sociocultural phenomena constricting their access to forest resources, women living in extreme poverty and dependent on forest resources are systemically ignored.

Despite the environmental policy, intervention has upgraded gender equity concerns over the last few years; women still remain impoverished by insecure access and constricted property rights to the forests, trees and land resources.[1] To illustrate, compared with men, women excessively tolerate the costs of tree and forest management, comprehend only a fraction of the benefits, and tend to be enlisted for decision-making only when forest and tree resources are ruined (Agrawal/Chhatre 2006). Besides lacking formal education, occupation and personal networks, rural women find themselves in the lowest position to influence resource allocation or research (Crewe/Harrison 1998; Ferrier 2002). One study demonstrates that changes in trees, and the loss of community access to forests, can have a disproportionately adverse impact on women, with an indirect impact on households, and consequently, on the livelihood of five to ten times as many people (Angelsen et al. 2012). It examines the interplay of power, institutions and practices that instigate disparities between men and women in tree and forest management.

2.3 Gender Relations in Forest-Related Research

Tyagi/Das (2018) argue that in recent years, gender has been a key element of study in forest policy and practice. Women's participation in forest governance is the most discussed topic in gender and forest research. Gender relations may play an important role in forest policies to achieve a sustainable socioecological outcome. Therefore, considering the gender relations among the people would be helpful for sound gender-responsive policies. Agarwal (2001) remarks that the knowledge system on forest resources is driven by a gendered division of work. Men do money-generating work (e.g. collecting wood for charcoal production), while women use forest resources for household works (e.g. collecting firewood and food) (Rocheleau et al. 1996). Research conducted in various seascapes of Unguja Island, Tanzania, suggests that there is a clear gendered division of labour and gender

---

[1]World Bank, FAO and IFAD 2009, *Gender in agriculture sourcebook*, Washington, DC.

symbolism (Torre-Castro et al. 2017). Women are responsible for reproduction activities. They remain in coastal forests and shallow areas. Women collect wood, invertebrates and farm seaweed. On the other hand, men use the whole seascape for their activities. Torre-Castro et al. (2017) found that men earn more money, which they call a 'universal pattern'. Women are stereotyped as 'housewives' only. Gender inequality was thus also present in seascape management.

Sunderland et al. (2014) hold that there is significant gender differentiation in the activities related to forest resources. It was found from the evidence of tropical Asia, Africa and Latin America that men have a more important and diverse role in the contribution of forest products. There is a difference between men's and women's knowledge, roles and skills in forest-related activities (Agarwal 2009; Bechtel 2010; Hecht 2007; Mai et al. 2011; Brown 2011; Rocheleau/Edmunds 1997). Previous research has suggested that there is often a similar gendered division of labour across cultures and regions (Bechtel 2010; Mai et al. 2011). For instance, men manage and use natural resources for harvesting, hunting, logging, construction and other money-generating activities (Agarwal 2009; Cavendish 2000; Shackleton et al. 2011; Shively 1997). On the other hand, women collect forest resources for household use (Agarwal 2009; Cavendish 2000), though women sometimes also use forest resources for commercial activities. Though men and women earn nearly the same amount from forest products globally, there are some regional variations. In Latin America, men contribute nearly seven times more than women to the family income from forest resources. Men are reported to earn slightly more than women in Asian countries. In contrast, women's contribution from forest resources is more than that of men in African sites.

Gender and forest researches consider the differences between men's and women's interactions, their particular roles, their knowledge in various forestry fields and the factors that reinforce any difference in evidence (Camou-Guerrero et al. 2007; FAO 2007; PRB 2001; USAID 2001 cited in Mai et al. 2011). Gender relations clearly describe the practical and strategic gender interests, the extent of dependence on forest resources (Agarwal 2010; FAO 2007), men's and women's relative priorities, access, control and power to make decisions over forest resources (Reeves/Baden 2000). Gender relations deal with a complex set of material and ideological aspects which is not limited to the division of labour and resources between men and women. Ideas and representations of women's and men's different abilities, attitudes, desires, personality traits and behavioural patterns are also the topics of study in gender relations (Agarwal 1997). Gender relations of a specific society are formed by these practices and ideologies of individuals (Agarwal 1997). The interactional aspects among individuals and social institutions shape the social stratification in a given context. Due to the unequal power relations between women and men in society, the natural resource management process is not social or gender neutral (Vernooy/Zhang 2006 quoted in Dhali 2009). This study further elaborates that the integration of gender analysis helps to develop a better understanding and awareness of the social and power relations that govern access, use and control over natural resources, and creates room for social actors (women and men) to manoeuvre and to enhance the bargaining and negotiating power of groups that are marginalized and discriminated against. It leads to empowerment and transformation, where they have more access to, control over and benefit from natural resources. Vernooy/Zhang (2006, quoted in Dhali 2009) conclude that indigenous women in the hilly regions of Bangladesh possess inadequate schooling about the techniques of using natural resources, leading them to face the gender-stereotyped treatment from their male counterparts.

Women's presence contributes to improved forest protection and better management. Engaging more women in forest and fishery decision-making bodies can bring a fruitful change in forest governance and conservation (Leisher et al. 2016). Bandiaky-Badji (2011) suggests that women's lack of representation in the local organizations in Senegal negatively affects food security. Vázquez-García/ Ortega-Ortega (2017) suggest two kinds of reasons why women should be included in forest governance. First, including women will increase governance efficiency. Second, this will guarantee the preservation of women's rights. FAO (2013) holds that women's engagement in land-related institutions can bring gender equity to land governance. If women are excluded from forest governance there may arise many negative outcomes (Ribot/Lee 2003). To improve the quality of natural resources and coastal management, gender should be considered critically. Better management cannot be achieved while the knowledge of women resource users is ignored and women's participation in decision-making processes is low (Torre-Castro et al. 2017).

2.4 Gender in the Environmental Security Discourse

Classmen (1995: 40, cited in Barnett 2011) argues that the environment and security relationships are an important association between resource scarcity and conflict. The dearth of environmental resources negatively affects human–environment relations. Although natural or man-made disasters befall the environment to its degradation, it is necessary to identify the underlying phenomenon causing the violent conflict between human settlements and the environment. The insufficiencies of resources threaten or destabilize the way of life of a given human population, or the internal structures of governance and activity through the fostering of subnational conflict and the significant reduction of options for action (Brunee/ Toope 1997). Furthermore, this study suggests redefining 'security', coined by Richard Ullman, outlining 'The sequence of events that: (a) threatens drastically and over a relatively brief span of time to degrade the quality of life for the inhabitants of a state, or (b) threatens significantly to narrow the policy choices available to the government of a state or to private, nongovernmental entities (persons, groups, corporations) within the state' (Ullman 1983, cited in Brunee/ Toope 1997: 26).

The scarcity of natural resources enforces a given society to adopt activities often subversive towards the environment. Barbier/Homer-Dixon (1997) divided the natural resources into two groups – renewables (forest, fresh water, land, fertile soil and Earth's ozone layer) and non-renewables (oil and minerals) – to describe how human beings consume resources for their own needs. This study illustrates that renewables are connected to human existence, but their scarcity severely undermines the well-being of people. For example, the degradation of the forest resources of a country might cause desertion. Dixon concludes his discussion by detecting three ways that weaken human well-being: a drop in the supply of a key resource, an increase in demand and a change in the relative access of different groups to the resource.

Detraz (2009) addresses the significance of incorporating gender into the approaches to the environment and security to exhibit the gender understanding of both security and the environment. The study analytically examines environmental security through a gender lens, clarifying the gendered nature of global environmental politics, and redefines the concept in ways that are more useful for understanding the juncture between environmental resources, gender relations and environmental security.

Detraz (2009) lists three distinctive viewpoints by merging security and environment – environmental conflict, environmental security and ecological security – and demonstrates how interactions between human beings and ecosystems affect the environmental security/ecological security.

The haphazard use of renewables by human beings for their survival, as mentioned by Dixon (1999), expedites environmental insecurity. The environmental degradation, either instigated by cyclones or man-made factors, threatens the ecosystem of a specific region.

Worldwide, forests serve an important role for the protection of the natural environment. The community-owned forest in the rural areas of Bangladesh saves people by giving them wood to be used as fuel and employment opportunities, whereas coastal forests and mangroves reduce the high speeds of winds and storms. Mangrove forests provide coastal erosion protection and preserve wetlands (Fritz et al. 2009). Forests play a significant role in controlling air pollution by absorbing pollutants such as sulphur dioxide, hydrogen fluoride and heavy metals, as well as ozone (Innes 2005). In addition, this study points out that forests protect global carbon cycles and thus represent an important sink for atmospheric carbon dioxide. The section 'gender in environmental security discourse' discusses the essence of environment and security in terms of resource scarcity and conflict induced by the human population. The further discussion takes gender lines into account and examines the traditional relationship between environment and security scholarship. Furthermore, the arguments summarize the implications of the forest for the lives of communities and show how the forest maintains the ecological balance in the coastal regions of Bangladesh.

2.5 Standpoint Theory

Standpoint theory emerged during the 1970s and 1980s from the women's liberation movement supported by a group of western feminist scholars like Nancy Hartsock (1983), Alison Jaggar (1989 quoted in Fuller 1999), Hilary Rose (1994 quoted in Fuller 1999), and Sandra Harding (1993, 2004a, b, 2008), as well as Neil Smith (1987). Among these scholars, Harding's argument is widely used in the social sciences. Initially, the theory borrowed its fundamental insights from Marx's historical materialism. Standpoint theory describes the underrepresented position of women of a specific social and cultural context. The term 'standpoint' is identical to 'perspective', which informs how knowledge is constructed and presented in a socially scientific manner in the academic text.

This theory seeks to explain the experiences of both women and men, and more specifically how women's experiences are shaped by social and political phenomena. The theory is applied in order to see beneath the ideological surface of social relations accepted as natural (Hartstock, cited in Harding 2006). Hekman (1997) argues that the original formulation of feminist standpoint theory is based on two major assumptions: all knowledge is located and situated. Hekman also discusses that location, the standpoint of women, is privileged, because it provides a vantage point in enunciating the truth of social reality.

Standpoint theory claims that some kinds of social locations and political struggles advance the growth of knowledge, contrary to the dominant view that politics and local situatedness can only block scientific inquiry (Harding 2004a, b: 351).

Knowledge is produced from the subjective positions of women's everyday lives. The experiences of the marginalized group of women are reflected and represented through systematic power relations. Therefore, standpoint theory discovers the dialectics of knowledge production based on the dominated versus the dominating groups, offering the *deconstructive* aspects in exposing the androcentric quirks within the theory and practice of the sciences and social sciences, and *reconstructive* ones in offering alternative explanations of the world informed by women's experiences and activities (Ho/Schraner 2004). It emphasizes that feminists and scholars ought to commence empirical and theoretical research projects 'from women's lives'. In this regard, the oppressed perspectives of women and their acquired experiences in daily lives are reflected through this theory. Therefore, Harding (2008: 115) addressed it as 'sciences from below'. As an organic epistemology, philosophy of science, social theory as well as methodology, the standpoint theory has received extensive coverage in many social as well as natural science disciplines (Harding 2008).

The essence of standpoint theory possesses four major dimensions in social science research projects (Harding 2004a, b). First, the motto is to articulate how the perspectives of women, or other socially relegated groups, are constituted, and to examine the practices of power of the dominant institutions (e.g. masculinity, family) and how their conceptual frameworks help to maintain oppressive social relations.

Second, taking material and political disadvantaged forms of oppression into account, the theory provides a distinctive insight into how a hierarchal social structure functions. In these views, Smith (1987) has outlined different ways in which women perform tasks and take responsibility for their daily lives, which makes them marginalized.

Third, it aims to record what women or members of oppressed groups actually say or believe in recognizing their social surroundings. Usually, the oppressed groups of a given society comprehend that their distorted representations of social relations are caused by dominant groups. In order to investigate the truth, standpoint theory as an organic epistemology suggests changing our minds about what our experiences were or how we want to think about marginalized groups. In this turn, the actual source of domination acting upon the marginalized group can be found through cross-checking their beliefs.

Fourth, standpoint theory concentrates more on the foundation of group consciousness rather than the shift in the individual's consciousness. Regarding this, Sandra Harding (2004a, b: 32) has expressed her position in the following way.

An oppressed group has to come to understand that each member is oppressed because she or he is a member of that group – Black, Jewish, women, poor, or lesbian – not because he or she individually deserves to be oppressed. The creation of group consciousness occurs (always and only?) through the liberatory political struggles it takes to get access to and arrive at the best conception of research for women or other oppressed groups, among the other goals of such struggles.

The context of the discovery of knowledge, and how it changes over time due to the influences of social actors (e.g. community, political party and NGOs), needs to be discussed. Harding (1993) argued this issue through the concept of 'Strong Objectivity' to explain women's subjective experiences attained from the objective phenomena. Hesse-Biber (2012) expanded Harding's work, describing the extent to which values and attitudes of the researcher also enter the 'context of discovery'. Feminist standpoint theorists contend that knowledge from the subordinated stratum is more complete than that of the dominant stratum. From this point of view, Mahatab (2010) argues that the rural women of Bangladesh comprehend their social world and the ways of survival in a critical situation better than men.

Epistemologically, are women of the less developed countries privileged agents in the context of knowledge production and their situatedness? Bringing 'Third World Women' into account, Mohanty (2003: 231, cited in Mjaaland 2013: 60) argues that feminist standpoint theory and epistemic privilege are understood with the analysis of experience, identity and the effect of social locations that can answer the issues of marginalization and use and abuse of power in the present transnational context. The experiences of women gained through their social world are shaped by the everyday political struggles. Standpoints of women in a given society are identified by their interactions with other women and men, as well as the resources they have access to in order to lead their lives. With regard to the perception of 'Third World women' as ignorant, poor, uneducated, traditionally bound, domestic family orientated and victimized; Mohanty (1991: 34–56) claims that the description of Third World feminists creates a discursive space where (self) knowledge is produced and the practice of recalling and representing hints at the founding of politicized awareness and self-identity. In addition, the author from her perspective of knowledge production defines it as an essential 'discursive site for struggle'.

The discursive arguments on standpoint theory in this study will analyse what standpoint forest dependents occupy in the study region and explore the perceptions of how forest resources are used. This investigation has been organized by observations in the study site over 3 months. The reflection of this scrutiny is discussed in the fifth chapter.

2.6 Feminist Political Ecology

Feminist political ecology is a bare-bones account of description and analysis of human–environment relations. It articulates the nexus between feminism and political ecology. In *Making Political Ecology*, Neumann (2005) notes that several scholars refer to political ecology as a research agenda (Bryant 1992, quoted in Neumann 2005), an approach (Warren et al. 200; Zimmerer/Bassett 2003a, quoted in Neumann 2005), and a perspective (Rocheleau et al. 1996; Kalipeni/Feder 1999, quoted in Neumann 2005). Beyond these, political ecology outlines how power relations and politics characterize the dynamics of economic development, environmental transformation and social change through a contextual geographic scale of analysis from the local to the global. As an approach, political ecology stands at the interdisciplinary crossroads of critical development studies, anthropology, feminism and environmental studies.

'Women and Life on Earth: A Conference on Eco-Feminism in the Eighties', held in March 1980, revealed the connections between feminism, militarization, healing and ecology. The conference described *Eco-feminism* as the connectedness and wholeness of the theory and practice. Mies/Shiva (1993) suggest that 'women have a deep and particular understanding of future generations and life on earth through their intimate connection between nature and experience'. Although *Eco-feminism* pointed out the connection between women and nature, it struggled to find out how men, as the counterpart of women, sustain the balance between nature and human setting. In response to the certain strands of eco-feminism, feminist political ecology has filled a need to theorize the role of gender without emphasizing the link between women and environments (Leach 2007). Men in the mangrove regions of Bangladesh usually go to the forest more than women. Before 2005 in the study site, few women would go to the forest, as their husbands earned the daily monetary support by selling forest resources at the local market.

Rocheleau et al. (1996) stated that feminist political ecology examines identity, difference and the significance of peoples' relation to sites of environmental change, degradation and struggles. Furthermore, Rocheleau and her colleagues have laid out three analytical themes: (1) gendered knowledge, (2) gendered environmental rights and responsibilities, and (3) gendered environmental politics and grass-roots activism. The defined first and third themes are important to this study and help to explain how women and men construct their insights into the forest and its essential resources. More specifically, they explain how women encounter local ideological and social barriers in accessing and using forest resources for their survival. Mollett/ Faria (2012) argue that feminist political ecology scholarship has re-emerged with a new energy, engaging with post-structural theory and acknowledging the role of spatial and embodied practices in constituting gender subjectivity.

Feminist political ecology as a theory proposes gender relations as an important marker of resource struggles scaled from the state to the body. Scholarly works (e.g. Mackenzie 1998; Gezon/Altamira 2006; Sultana 2011; Truelove 2011) pay close attention to struggles over household resources, gender-wise division of labour and livelihood security, as activities concerning those phenomena are unfolded in everyday practices and engender body politics. The struggle for environmental resources between women and men in the household, or even in the public space, is shaped by conventional community politics. Mollet/Faria (2012) further suggest that feminist political ecology focuses on gender and household relations in providing a nuanced conceptualization of gender relations in the context of development interventions.

In her work 'Gendered livelihoods and the politics of socio-environmental identity: women's participation in conservation projects in Calakmul, Mexico' Radel (2012), points out the interaction between the social construction of people's relation to their environments and the social construction of gender. In stressing the perspective of socio-environmental identities,[2] the writer explained that social constructions of people's relations in a specific sociopolitical context of environmental resources are determined through their labour, their livelihoods and their environmental ethics. Radel's thoughts of gendered ideologies of natural resource access, control and decision-making offer an essential link between feminist political ecology and political ecology that examines how resource rights are negotiated between men and women within both households and villages. She states:

> …in investigating the link between gender relations and environmental relations, there has been an increasing emphasis in feminist political ecology on mutual construction, with a particular stress on the importance of the ideological, including identity, in concert with the material, in the creation of gendered environmental relations (Gururani 2002; Nightingale 2006; Rocheleau et al. 2001, cited in Radel 2012: 66).

---

[2]Radel (2012) has employed the concept of identity to refer to the dual construction of the individual both in terms of the individual's sense of self and in terms of the labelling of the individual by others or by society. Identities should be thought of as shifting, contingent and relational (Haraway 1991; Harding 1998, cited in Radel 2012), and are both constructed and performed through ordinary, daily practices (Butler 1990; McDowell 1995, cited in Radel 2012).

Radel's theoretical position fits in the context of the study, as women informants are treated inhumanely in the study setting for their forest-going practices. Even the local religious leaders advise men and young people not to interact with *divorced women*. This aspect of negative labelling of women questions their socio-environmental identities, as claimed by the author. The insights of the existing body of literature (e.g. Gururani 2002; Bondi 2005; Pile 2010) into feminist political ecology direct the ways that emotions come to matter in nature–society relations and influence why and how people use, gain access to, control and conflict over the ways they do.

Sultana (2011) has outlined feminist political ecology, mentioning the messiness of everyday politics and struggles over a critical resource, such as water. In grounding her framework, she confesses that ideas of access, use and control of resources are intertwined and interconnected within the embodied emotions critical to explaining the ways that nature–society relationships operate in everyday life in any given context. Thus, the fragile and tough process involved in accessing a necessary resource (forest) poses material and logistic challenges for women in the study region. The proximity, distance between home, source of the resource, time needed, gendered space and physical burdens determine how men and women, particularly women, overcome such challenges.

Radel's argument has been reflected in Sultana's (2011) work on drinking water contamination in Bangladesh, which articulates the importance of heeding the various emotions and meanings attached to processes of resource access, use and conflict in order to better understand the emotionality of the resources existing in everyday struggles. These analyses enable feminist political ecology to clarify and illuminate the ways that resources struggles and politics are not only economistic, social or rational choice issues, but also emotive realities which have a direct bearing on how resources are accessed, used and fought over.

Elmhirst (2018) has pointed out that the cross-border migration between East Kalimantan and Malaysia related to oil palm labour markets is very significant. The labourers who are returning are investing their skills, capital and tools in their locality. According to Collier and Ong (2005), material feminism has a view that the materiality of the natural world has a complex relationship with the natural elements. Peluso has successfully explained 'mobile and material feminist political ecology' (Elmhirst 2018). The literature on oil palm agriculture suggests that there might be a wide number of gendered impacts and responses (Elmhirst et al. 2017). According to Park/Daley (2015), gender dimensions of land acquisitions remark that social relations that include generation connectivity play important roles in gender dynamics. In terms of access to forest resources and forest-based livelihoods, 'gender' and 'gender inequality' must be studied with reference to context. In Indonesia, gender is not given as much importance as other life phenomena (Sears 1996). Sometimes, gender is realized based on its relation with ethnicity and customs (locally called 'adat'). The Indonesian state itself and its rules also play roles in shaping gender norms (e.g. males are considered breadwinners and females caregivers). There also exist gender divisions of labour in oil palm and plantation. Males do the harder tasks, while women do easier tasks (Elmhirst et al. 2017).

Research by Elmhirst et al. (2017) has shown how gendered and generational experiences in context are formed through landscape history and modes of incorporation into oil palm systems. Oil palm has played a critical role in reinforcing contrasts amongst differently positioned actors and communities. Moreover, many gender and intergenerational norms are so entrenched that fixity hampers change in many sectors. According to Elmhirst et al. (2017), big investments in oil palm can affect the situation. In the study area, it was found that women get less opportunity at institutional level because of the established family and socially gendered norms. Thus, they stay away from the decision-making process.

Lamb et al. (2017) argued that there are many unresolved issues related to gender and land-grabbing in Cambodia. The research identified the relationship between gender and race, ethnicity, indigeneity and class (Lamb et al. 2017). The issues of educational opportunities with reference to class and gender should be addressed in Cambodia. Nyantakyi-Frimpong (2017) holds that gendered politics have a huge impact on access to food, which shapes agricultural diversification. Agricultural diversification contributes to dietary diversity. There are structural constraints limiting women's access to resources. To understand gender, land and the state well, research on masculinity in Asia may be useful. Research should consider both men and women to bring out a better outcome in gender relations (Lamb et al. 2017).

To conclude, the framework of Rocheleau et al. (1996) has been used to analyse the gendered concerns relating to environmental resources with particular regard to forest resources. The recent argument produced by Radel (2012) has been used to understand how the social construction of people's relation to their environments and the social construction of gender are formed in the study context. Meanwhile, Sultana's sketch of the feminist political ecology framework in the context of Bangladesh reveals the daily politics and struggles which increase the burden of women's activities both within the forest and at home.

2.7 Contribution of the Study

This chapter addresses how gender lines are considered in the discourse of forest and environmental security, but also shows that the discussion seldom maps out the forestry society's interactional aspects to the mangrove forest. It further points out how the human–environment relationship is impacted with regard to cyclones in the mangrove regions of Bangladesh. Moreover, this chapter also discussed standpoint theory by focusing on the argument of Hekman (1997) as well as several works of Harding. The presentation of scholars' insights into standpoint theory demonstrates the theoretical orientation to be applied for identifying women's and men's standpoints with regard to forest-connected activities. Feminist political ecology theory is employed to analyse how women have access to the forest and gain control over it, and their behaviour towards the forest and its collected resources. However, the literature review hardly makes evident the role of forests in reducing wind velocity of cyclones as well as women's and men's indigenous knowledge of the forest and ecological protection. Therefore, the knowledge gap which has emerged from the literature review offers an opportunity to investigate from a fresh perspective the indigenous perceptions of women and men of the Sundarbans forest, how they behave towards the forest, and the interconnection between the forest and ecological security. As mentioned in Chapter One, this study uses empirical data and the findings gathered from fieldwork to fill this knowledge gap. The next chapters will discuss these findings.

References

Agarwal, B. (1997), '"Bargaining" and Gender Relations: Within and Beyond the Household', *Feminist Economics*, Vol. 3, No. 1, pp. 1–51.

Agarwal, B. (2001), 'Participatory Exclusions, Community Forestry, and Gender: An Analysis for South Asia and A Conceptual Framework', *World Development*, Vol. 29, No. 10, pp. 1,623–1,648.

Agarwal, B. (2009), 'Gender and Forest Conservation: The Impact of Women's Participation in Community Forest Governance', *Ecological Economics*, Vol. 68, No. 11, pp. 2,785–2,799.

Agarwal, B. (2010), 'Does Women's Proportional Strength Affect their Participation? Governing Local Forests in South Asia', *World Development*, Vol. 38, No. 1, pp. 98–112.

Agrawal, A. and Chhatre, A. (2006), 'Explaining Success on the Commons: Community Forest Governance in the Indian Himalaya', *World Development*, Vol. 3, No. 1, pp. 149–166.

Alam, M.J.; Rakkibu, M.G.; Rahman, M.M. (2012), 'People's Attitudes Towards Social Forestry: A Case Study in Rajshai', *Environmental and Natural Resources*, Vol. 5, No. 1, pp. 217–222.

Angelsen, A.; Brockhaus, M.; Sunderlin, W.D.; Verchot, L.V. (eds.) (2012), *Analysing REDD+: Challenges and Choices*, CIFOR, Bogor, Indonesia.

Bandiaky-Badji, S. (2011), 'Gender Equity in Senegal's Forest Governance History: Why Policy and Representation Matter', *International Forestry Review*, Vol. 13, No. 2, pp. 177–194.

Barbier, E. and Homer-Dixon, T. (1997), 'Resource Scarcity and Innovation: Can Poor Countries Attain Endogenous Growth?', *Ambio*, Vol. 28, No. 2, pp. 144–147.

Barnett, J. (2011), *The Meaning of Environmental Security: Ecological Politics and Policy in the New Security Era*, London: Zed Books Limited.

Belcher, B.M.; Ruiz-Perez, M.; Achdiawan, R. (2005), 'Global Patterns and Trends in the Use and Management of Commercial NTFPs: Implications for Livelihoods and Conservation', *World Development*, Vol. 33, No. 9, pp. 1,435–1,452.

Bose, P, (2011), 'Forest Tenure Reform: Exclusion of Tribal Women's Rights in Semi-Arid Rajasthan, India', *International Forestry Review*, Vol. 13, No. 2, pp. 220–232.

Bosold, L.A. (2012), 'Challenging The "Man" In Mangroves: The Missing Role of Women in Mangrove Conservation', *Student Publications*, Paper 14, viewed 19 January 2013, http://cupola.gettysburg.edu/student_scholarship/14.

Bondi, L. (2005), 'Making Connections and Thinking through Emotions: Between Geography and Psychotherapy', *TIBG*, Vol. 30, No. 4, pp. 433–448. https://doi.org/10.1111/j.1475-5661.2005.00183.x.

Brown, P.C. (2011), 'Gender, Climate and REDD+ in the Congo Basin Forests of Central Africa', *International Forestry Review*, Vol. 13, No. 2, pp. 163–176.

Brunnée, J. and Toope, S. (1997), 'Environmental Security and Freshwater Resources: Ecosystem Regime Building', *American Journal of International Law*, Vol. 91, No. 1, pp. 26–59. https://doi.org/10.2307/2954139.

Cannon, T. (2002), "Gender and Climate Hazards in Bangladesh", *Gender & Development*, Vol. 10, No. 2, pp. 45–50. https://doi.org/10.1080/13552070215906.

Cavendish, W. (2000), 'Empirical Regularities in the Poverty-Environment Relationship of Rural Households: Evidence from Zimbabwe', *World Development*, Vol. 28, pp. 1,979–2,003.

Collier, S.J. and Ong, A. (2005), 'Global Assemblages, Anthropological Problems'. In: A. Ong and S.J. Collier (eds.), *Global Assemblages: Technology, Politics and Ethics as Anthropological Problems*, Oxford: Blackwell, pp. 3–21.

Crewe, E. and Harrison, E. (1998), *Whose Development? An Ethnography of Aid*, London: Zed Books.

Detraz, N. (2009), 'Environmental Security and Gender: Necessary Shifts in an Evolving Debate', *Security Studies*, Vol. 18, No. 2, pp. 345–369.

Dixon, T.F.H. (1999), 'Environment, Scarcity and Violence', *Human Ecology Review*, Vol. 7, No. 1, Princeton, NJ: Princeton University Press.

Dhali, H.H. (2009), *'Sustainable Development, State Policy, and Gender: Examining the effect of the forest policy on gender relations of indigenous people in CHT in Bangladesh'*, The Hague: International Institute of Social Studies (I.S.S.), pp. 1–58.

Elmhirst, R.; Siscawati, M.; Basnett, B.S.; Ekowati, D. (2017), 'Gender and Generation in Engagements with Oil Palm in East Kalimantan, Indonesia: Insights from Feminist Political Ecology', *The Journal of Peasant Studies*, Vol. 44, No. 6, pp. 1,135–1,157. https://doi.org/10.1080/03066150.2017.1337002.

Elmhirst, R. (2018), 'Understories of the Political Forest: A Mobile Feminist Political Ecology? Commentary on Nancy L. Peluso's "The Remittance Forest: Turning Mobile Labour into Agrarian Capital"', *Singapore Journal of Tropical Geography*, Vol. 39, pp. 41–44.

FAO (2011), *The State of Forests in the Amazon Basin, Congo Basin and Southeast Asia*, Rome: United Nations Food and Agriculture Organization. http://www.fao.org/docrep/014/i2247e/i2247e00.pdf.

Ferrier, S. (2002), 'Mapping Spatial Pattern in Biodiversity for Regional Conservation Planning: Where to from Here?', *Systematic Biology*, Vol. 51, No. 2, pp. 331–363.

Fritz, T.; Jentschke, S.; Gosselin, N.; Sammler, D.; Peretz, I.; Turner, R.; Friederici, A.D.; Koelsch, S. (2009), 'Universal Recognition of Three Basic Emotions in Music', *Current Biology*, Vol. 19, No. 7, pp. 573–576. https://www.ncbi.nlm.nih.gov/pubmed/19303300.

Fuller, D. (1999), 'Helen Porter's Everyday Survival Stories: A Literary Encounter with Feminist Standpoint Theory', *Atlantics*, Vol. 24, No. 1, pp. 122–133.

Gezon, L.L. (2006), *Global Visions, Local Landscapes: A Political Ecology of Conservation, Conflict, and Control in Northern Madagascar*, Lanham, MD: Rowman & Littlefield AltaMira Press.

Giri, K. (2012), *'Gender in Forest Tenure: Pre-requisite for Sustainable Forest Management in Nepal'*, Washington DC: Rights and Resources.

Gurarascio, F. et al. (2013), *Forests, Food Security and Gender: Linkages, Disparities and Priorities for Action* (Background paper for the International Conference on Forests for Food Security and Nutrition), Rome: United Nations Food and Agriculture Organization, http://www.fao.org/docrep/018/mg488e/mg488e.pdf.

Gururani, S. (2002), 'Construction of Third World Women's Knowledge in the Development Discourse', *International Social Science Journal*, Vol. 54, No. 154, pp. 313–323.

Halim, S. (1999), *'Invisible Again: Women and Social Forestry in Bangladesh'*, Ph.D. Thesis, McGill University, Montreal, Canada.

Harding, S. (2004a), 'A Socially Relevant Philosophy of Science? Resources from Standpoint Theory's Controversiality', *Hypatia*, Vol. 19, No. 1, pp. 25–47.

Harding, S. (2004b), *The Feminist Standpoint Reader* (ed.), New York and London: Routledge.

Harding, S. (2008), *Sciences from Below: Feminisms, Postcolonialities, and Modernities*, Durham, NC: Duke University Press.

Hartsock, N. (1983), 'The Feminist Standpoint: Developing the Ground for a Specifically Feminist Historical Materialism'. In: Sandra Harding (ed.) *Feminism and Methodology: Social Science Issues*, Bloomington: Indiana University Press, pp. 157–180.

Hecht, S. (2007), 'Factories, Forests, Fields and Family: Gender and Neoliberalism in Extractive Reserves', *Journal of Agrarian Change*, Vol. 7, No. 3, pp. 316–347.

Hekman, S. (1997), 'Truth and Method: Feminist Standpoint Theory Revisited', *Chicago Journals*, Vol. 22, No. 2, pp. 351–365.

Hesse-Biber, S.N. (ed.) (2012), *Handbook of Feminist Research Theory and Praxis*, Boston, MA: SAGE Publications.

Hossain, M.M. and Islam, M.M. (2017), 'Community Dependency on the Ecosystem Services from the Sundarbans Mangrove Wetland in Bangladesh', *Wetland Science*, pp. 301–316. https://doi.org/10.1007/978-81-322-3715-0_16.

Ho, C. and Schraner, I. (2004), *Feminists Standpoints, Knowledge and Truth*, Paper No. 2004/02, School of Economics and Finance, University of Western Sydney, Australia.

Iftekhar, M.S. and Islam, M.R. (2004), 'Managing Mangroves in Bangladesh: A Strategy Analysis', *Journal of Coastal Conservation*, Vol. 10, pp. 139–146.

Innes, J.L. (2005), 'Forests in Environmental Protection', *Forests and Plants*, Vol. 1, pp. 1–7.

Kabir, H.M.D. and Hossain, J. (2008), *Resuscitating the Sundarbans, Customary Use of Biodiversity and Traditional Practices in Bangladesh*, Dhaka, Bangladesh: Unnayan Onneshan; The Innovators.

Leisher, C.; Temsah, G.; Booker, F.; Day, M.; Samberg, L.; Prosnitz, D.; Agarwal, B.; Matthews, E.; Roe, D.; Russel, D.; Sunderland, T.; Wilkie, D. (2016) 'Does the Gender Composition of Forest and Fishery Management Groups Affect Resource Governance and Conservation Outcomes?: A Systematic Map', *Environmental Evidence*, Vol. 5, No. 6. https://doi.org/10.1186/s13750-016-0057-8.

Lamb, V.; Schoenberger, L.; Middleton, C.; Un, B. (2017), 'Gendered Eviction, Protest and Recovery: A Feminist Political Ecology Engagement with Land Grabbing in Rural Cambodia', *The Journal of Peasant Studies*. https://doi.org/10.1080/03066150.2017.1311868.

Leach, M. (2007), 'Earth Mother Myths and Other Ecofeminist Fables: How a Strategic Notion Rose and Fell', *Development and Change*, Vol. 38, No. 1, pp. 67–85.

Mackenzie, C. (1998), 'The Choice of Criteria in Ethical Investment', *Business Ethics*, Vol. 7, No. 2. pp. 81–86. https://doi.org/10.1111/1467-8608.00092.

Mahatb, N. (2010), *Women in Bangladesh*, Dhaka: Dhaka University Press.

Mai, Y.H.; Mwangi, E.; Wan, M; (2011), 'Gender Analysis in Forestry Research: Looking Back and Thinking Ahead', *International Forestry Review*, Vol. 13, No. 2, pp. 245–258.

McDougall, C.L.; Leeuwis, C.; Bhattarai, T.; Maharjan, M.R.; Jiggins, J. (2013), 'Engaging Women and the Poor: Adaptive Collaborative Governance of Community Forests in Nepal', *Agriculture and Human Values*, Vol. 30, pp. 569–585. https://doi.org/10.1007/s10460-013-9434-x.

Mjaaland, T. (2013), 'At the Frontiers of Changes? Women and Girls' Pursuit of Education in North-Western Tigray, Ethiopia', Ph.D. thesis, Bergen: University of Bergen.

Mies, M. and Shiva, V. (1993), *Ecofeminism*, London and New Jersey: Zed Books.

Mohanty, T.C. (2003), *Feminism without Borders*, Durham, NC: Duke University Press.

Mollett, S. and Faria, C. (2012), 'Messing with Gender in Feminist Political Ecology', *Geoforum*, viewed 1 February 2013, http://dx.doi.org/10.1016/j.geoforum.2012.10.009.

Mukasa, C.; Tibazalika, A.; Mango, A.; Muloki, H.N. (2012), *Gender and Forestry in Uganda: Policy, Legal and Institutional Frameworks*, Bogor, Indonesia: CIFOR.

Nasrin, F. (2012), 'Women, Environment and Environmental Advocacy: Challenges for Bangladesh', *Asian Journal for Social Sciences and Humanities*, Vol. 1, No 3, pp. 149–172.

Neumann, P.R. (2005), 'Introduction', in: P.R. Neumann (ed.), *Making Political Ecology*, London and New York: Routledge.

Nyantakyi-Frimpong, H. (2017), 'Agricultural Diversification and Dietary Diversity: A Feminist Political Ecology of the Everyday Experiences of Landless and Smallholder Households in Northern Ghana', *Geoforum*, Vol. 86, pp. 63–75.

Park, C.M.Y. and Daley, E. (2015), 'Gender, Land and Agricultural Investments in Lao PDR', in: C.S. Archambault and A. Zoomers (eds.), *Global Trends in Land Tenure Reform: Gender Impacts*, London: Routledge.

Pile, S. (2010), 'Emotions and Affect in Recent Human Geography', *TIBG*, Vol. 35, No. 1, pp. 5–20. https://doi.org/10.1111/j.1475-5661.2009.00368.x.

Radel, C. (2012), 'Gendered Livelihoods and the Politics of Socio-Environmental Identity: Women's Participation in Conservation Projects in Calakmul, Mexico', *Gender, Place & Culture: A Journal of Feminist Geography*, Vol. 19, No. 1, pp. 61–82.

Rocheleau, D. and Edmunds, D. (1997), 'Women, Men and Trees: Gender, Power and Poverty in Forest and Agrarian Landscapes', *World Development*, Vol. 25, No. 8, pp. 1,351–1,371.

Rocheleau, D.; Thomas-Slayter, B.; Wangari, E. (1996), 'Gender and Environment: A Feminist Political Ecology Perspective', in: D. Rocheleau, B. Thomas-Slayter, and E. Wangari (eds.), *Feminist Political Ecology: Global Issues and Local Experiences*, London and New York: Routledge, pp. 3–23.

Ribot, J. and Lee, N.P. (2003), 'A Theory of Access', *Rural Sociology*, Vol. 68, No. 2, pp. 153–181.

Sears, L.J. (1996), *Fantasizing the Feminine in Indonesia*, Durham, NC: Duke University Press.

Sen, A. (1990), *Development as Freedom*, New York: Knopf Doubleday Publishing Group.

Shanley, P.; Silva, F.C.D.; MacDonald, T. (2011), 'Brazil's Social Movement, Women and Forests: A Case Study from the National Council of Rubber Tappers', *International Forestry Review*, Vol. 13, No. 2, pp. 233–244.

Shackleton, S.; Paumgarten, F.; Kassa, H.; Husselman, M.; Zida, M. (2011), 'Opportunities for Enhancing Poor Women's Economic Empowerment in the Value Chains of Three African Non-timber Forest Products (NTFPs)', *International Forestry Review*, Vol. 13, No. 2, pp. 136–151.

Shackleton, C.M. and Shackleton, S.E. (2000), 'Direct-Use Value of Secondary Resources Harvested from Communal Savannas in the Bushbuckridge Low Veld, South Africa', *Journal of Tropical Forest Products*, Vol. 6, pp. 28–47.

Shively, G.E. (1997), 'Poverty, Technology, and Wildlife Hunting in Palawan', *Environmental Conservation*, Vol. 24, No. 1, pp. 57–63.

Siar, S.V. (2003), 'Knowledge, Gender, and Resources in Small-Scale Fishing: The Case of Honda Bay, Palawan, Philippines', *Environmental Management*, Vol. 31, No. 5, pp. 569–580. https://doi.org/10.1007/s00267-002-2872-7.

Smith, N. (1987), 'Gentrification and the Rent Gap', *Annals of the Association of American Geographers*, Vol. 77, No. 3, pp. 462–465. https://doi.org/10.1111/j.1467-8306.1987.tb00171.x.

Stiem, L. and Krause, T. (2016), 'Exploring the Impact of Social Norms and Perceptions on Women's Participation in Customary Forest and Land Governance in the Democratic Republic of Congo – Implications for REDD+', *International Forestry Review*, Vol. 18, No. 1, pp. 110–122, Commonwealth Forestry Association. http://www.bioone.org/doi/full/10.1505/146554816818206113.

Sultana, F. (2011), 'Suffering from Water, Suffering for Water: Emotional Geographies of Resource Access Control and Conflict,' *Geoforum*, Vol. 42, No. 2, pp. 163–172.

Sun, Y.; Mwangi, E.; Dick, M.R.; Bose, P.; Shanley, P.; Silva, C.F.; MacDonald, T. (2012), 'Forests: Gender, Property Rights and Access', No. 47, Bogor, Indonesia: Center for International Forestry Research.

Sunderland, T.; Achdiawan, R.; Angelsen, A.; Babigumira, R.; Ickowitz, A.; Paumgarten, F.; Reyes-García, V.; Shively, G. (2014), 'Challenging Perceptions About Men, Women, and Forest Product Use: A Global Comparative Study', *World Development*, Vol. 64, No. 1, pp. S56–S66. https://doi.org/10.1016/j.worlddev.2014.03.003.

Tanny, N.Z.; Rahman, M.W.; Ali, R.N. (2017), 'Climate-induced Gender Vulnerabilities in Northwestern Bangladesh', *Indian Journal of Gender Studies*, Vol. 24, No. 3, pp. 360–372.

Torre-Castro, M.; Fröcklin, S.; Börjesson, S.; Okupnik, J.; Jiddawi, N.S. (2017), 'Gender Analysis for Better Coastal Management – Increasing Our Understanding of Social-Ecological Seascapes, *Marine Policy*, Vol. 83, pp. 62–74.

Truelove, Y. (2011), '(Re-) Conceptualizing Water Inequality in Delhi, India Through a Feminist Political Ecology Framework', *Geoforum*, Vol. 42, No. 2, pp. 143–152.

Tyagi, N. and Das, S. (2018), 'Assessing Gender Responsiveness of Forest Policies in India', *Forest Policy and Economics*, Vol. 92, pp. 160–168.

Vázquez-García, V. and Ortega-Ortega, T. (2017), 'Gender, Local Governance and Non-timber Forest Products. The Use and Management of Satureja Macrostema in Oaxaca's Central Valleys, Mexico', *Women's Studies International Forum*, Vol. 65, pp. 47–52.

Chapter 3
Narratives of the Sundarbans Forest at Shora

Abstract This chapter explores the lives of the Shora community in relation to the Sundarbans forest. The people of Shora grow and live with the Sundarbans, as it is the main source of their livelihoods and a core part of their lives. This chapter describes the empirical findings that show how the people of Shora have developed indigenous and customary knowledge about the Sundarbans and the collection of forest resources for several generations. Whilst women and men collect forest resources differently and with different environmental impacts, there is also consideration of the illegal activities in the forest that are causing long-term environmental insecurity. This chapter also acknowledges that there has been a noticeable change in the widely used customary knowledge system of the Shora community and concludes that the passing on of knowledge from one generation to the next is helping retain the individual knowledge system.

Keywords Sundarbans · Shora · Knowledge of the forest · Forest resources

3.1 Introduction

The local population of Shora dependent on the forest has developed an indigenous knowledge system over many years, and as the pattern is discernible in the overall perceptions of the participants about the region, its resources and usage, this chapter intends to decode the structure of this knowledge. To make sense of this knowledge, it is categorized into initial knowledge acquisition, practice of forest ranging and procurement of services from forest resources. By incorporating pertinent literature, this chapter uses standpoint theory, focuses on experiences of forest rangers about their use of natural resources and reviews previous studies of the region.

© The Author(s), under exclusive license to Springer Nature Singapore Pte Ltd. 2019 43
S. Roy, *Climate Change Impacts on Gender Relations in Bangladesh*,
SpringerBriefs in Environment, Security, Development and Peace 29,
https://doi.org/10.1007/978-981-13-6776-2_3

3.2   Acquaintanceship with the Forest

The observation data of the local population mud houses indicates extreme poverty and the penurious conditions in which they eke out a living in the struggle to survive. This phenomenon transcends the boundary of the mundane struggle of life and encroaches into the entertaining and didactic rituals of storytelling. Such stories, narrated by the senior members of the family such as parents, grandfathers and grandmothers, describe the tragic scenarios of their lives in relation to the forest. Turning into a trend over the years in the locality, storytelling occurs in the time slot of late evenings after dinner.

During summer rains, those who go to the forest stay inside their houses, spending leisure time conversing with family members, telling stories about the history of Sundarbans, long-established forest practices and beliefs, how forbearers used the forest and the present-day challenges. Generations hand down the stories and the practice of telling them. During winter, participants endure intense cold weather due to the wind blowing from the river *Kholpatura*. On wintry nights, family members of each household congregate in the courtyard to keep themselves warm by lighting a fire with leaves, straw and wood. As neighbours also join the host to warm by the fire, this creates an interactive environment where participants converse with each other about the daily happenings of the household. The discussion touches on collected forest resources and the market value of the resources, and they make predictions as to how long the market values can sustain the family livelihood. Listening to such conversations provides a clear indigenous picture of the forest.

Participant: 'Since 1975, I have lived in this village. After the sad demise of my father, my widowed mother nurtured me. The best guide in my life, she taught me what to do and what not to. She died in the devastating cyclone Aila in 2009. When I was seven, I once could not help but cry at the prospect of a violent tempest. The speed of the wind was horrifying, so I was unable to sleep that night. My mom came to me, caressed my head gently, and tried assuaging my fear, saying that Sundarbans, the almighty God, would protect us from the wrath of this tempest. The next morning, I woke up to find our house and our neighbours' inviolate. From that rainy night, I came to understand that God infused the forest with great significance for our life'. (A female participant, age 38.)

Participant: 'Men in childhood become acquainted with the forest with the help of their parents, in-laws, and next-door neighbours. As soon as we turn 20, 90% of men of this area go to the forest with their father, and come to know what are the important types of plants and trees, and how trees and timbers are used in the household and sold in the market'. (A male participant, age 35.)

Such stories demonstrate how storytelling is the primary step in building children's acquaintanceship with the forest, and in the study setting, the growing youth community prefers to interact with *Murubee*, one of the senior members of the community, aged over 70, who is better acquainted with the forest. They may share experiences either sitting at the tea stall or on the riverbank adjoining the forest.

A *Murubee* is revered; pious, experienced and respectable, they are part of the Shora think tank due to their possession of indigenous knowledge, life and the environment. *Murubees* like to call the Sundarbans *Bada*, *Mal* or jungle, and one *Murubee* man says, 'These local terms are frequently spoken by the area inhabitants for years'. During observation on a sunny evening, the *Murubee* enthusiastically shared his experiences of acquaintanceship with the Sundarbans upon the request of the researcher.

Participant: 'The Bada is like my son and daughter. Our existence is almost inconceivable without it. We had a joint family, together with my grandfather and grandmother. My father was the sole earner who used to go to the forest regularly. Once, when our home was at risk, with the permission of the community leader my father made a small house near the riverbank in April 1970. I recall that by shifting straw, wood, and timber of Bada (sic), I helped my father and uncles build the house. When it was completed, my grandfather told me that my father dedicated his life to managing food and belongings for the family members. It was a challenging task to collect forest resources from the Bada. I was encouraged to go to the forest with my father so that he might be supported. As I was a kid, I could not satisfy what he expected of me. The wise grandfather once managed a boat for me to visit the Bada with him. As we were canoeing in the river furrowing through the forest, he told me a story chronicling how the Bada has become Sundarbans'. (A male participant, age 73.)

The in-depth interviews and focus group data exploring the women's acquaintanceship with the Sundarbans finds that the Shora women are normally shy with people they do not know and may feel shy speaking to their husbands in the presence of others. Even when their husbands told them to speak out, their words – compared with those of the men – betrayed their hesitation to share their insights into the forest.

Participant: 'In 1993, when I was in Class Five (Grade 5) at school in the forest region, I read a chapter on the importance of forest in Bangladesh in the social studies course. Before going through the chapter, the teacher brought all the class participants outside the classroom. He told the students to look at the forest; he then described the beautiful green forest and explained the necessity of its resources for lives. It was a little confusing for me to believe the words of my teacher. Not understanding what roles Bada plays for us, I felt the need to bring the matter to my father, a honey-collector. After ranging into the heart of the forest for more than two weeks accompanied by some of his colleagues, he returned, drums replete with honey; it made all of us elated. Over dinner, he recounted how honey was collected. Startled to hear from him, I could justify the words learnt from my teacher. Though I never visited Bada, after getting married I listened to a similar story from my husband'. (A female participant, age 54.)

Female participants become familiar with the forest from the lessons they learn at primary school and from listening to their fathers' stories and their husbands' descriptions. However, as the core duty of wives is to take care of their husbands and other family members, the wives' social work is restricted to the domestic sphere. The female participants said that in the absence of their husbands, they like to chat with the neighbouring women about the forest resources collected by their husbands. This has become a tradition exercised over the years in the region, as several of the women interviewed observed. Their opinions about the forest are influenced by their subservient attitude to the patriarchal agent, a phenomenon with a resemblance to male supremacy as an accepted part of the natural order of the patriarchal society of Bangladesh, where women's ideological stance in the community level is affected by the men's gender ideology (Sultana 2010). The traditional Bangladeshi society understands women as daughters, wives and mothers. This determines their role as the caregiver and household manager, while men, as the 'head' of the family, are deemed to be the financial providers. In accordance with Sultana's observations (2010), women's gender ideology against the backdrop of patriarchy is strongly supported by evidence from the above discussions, when women only spoke about the forest after receiving permission from their husbands.

3.3 Forest-Going Practice

The observations reveal that women living with their husbands hardly have access to the forest, because their husbands earn money for the family. For women without husbands, the study documents *Jele-Baoalie* and *Divorced Women* who have access to the forest for their livelihood, as opposed to the married women, who only rarely venture outside the home. The *Jele-Baoalie* and *Divorced Women* are socially vulnerable in the region, as they need to venture into the forest for survival. The study also documents the characteristics that proliferate amongst forest-ranging men and aims to spotlight the forest-going practice of the participants. This includes evolving long-established beliefs as an offshoot of such ventures, resulting in interpreting the theoretical features of standpoint theory.

3.3.1 Jele-Baoalie Women

Jele-Baoalie women are the fishing community in the forest, who fish all year round and make fishing nets using cord, colours and pieces of iron imported from the Indian market. The empirical data reveals that the *Jele-Baoalie* women are members of the Muslim community at Shora and, even before the morning breaks, they set off on their journey from the south of Shora towards the Sundarbans. In groups of five or six, the women board a wooden boat to travel one and a half hours into the forest. They alight from the boat when the water is chest-high and catch a variety of fish, especially the *pona* (fingerlings) from the canals of the deep forest, using a special kind of blue net (e.g. *ber jal, net jal, box jal*). Each net is placed in the water for a while so the fish are trapped, and then it is pulled from one side to another (e.g. left to right) from the canal-side. After pulling for approximately 15 min on one side, the women change direction and resume dragging it for another 15 min on the opposite side. The women usually fish in the canal bed with their hands to frighten the *pona* which are hiding in the riverbed. This makes the fish to come out and the net traps them.

The *Jele-Baoalie* possesses indigenous knowledge of catching *pona* called *ochol*. *Ochol jhara or ochol deya* involves collecting *pona* from the net and sorting *pona* from the unwanted, accumulated dirt. In the *ochol*, the practice separates *pona* from the crabs, *meku* and *methi*. The women use a transparent snail shell to collect the black *pona*, including the young fish, *bagda* and *chati*, and put them in a silver pot filled with water. Besides catching fish, the women collect firewood using a chopper to cut dead trees from the river and canal-side areas. In so doing, they follow two indigenous patterns of cutting trees. The first is *kolomchekur*, similar to a fountain pen and the second is *ar cope*, a plain style of tree cutting. The women also collected small branches and leaves from the forest.

According to the 1993 legislation adopted by the Ministry of Education of Bangladesh, there are no fees for primary schooling up to Grade 5. Although this law, irrespective of caste, creeds, gender and religion inspires the grass-roots children to go to school, the *Jele-Baoalie* women's children are not in a position to take the opportunity. Born into extremely poor families, largely ignored by the relatively rich in the community and typically construed as a domestic servant after marriage, this binds *Jele-Baoalie* women in the community. Confined to the acceptable areas of childbirth, child-rearing, serving the other family members and foraging in the forest for fish, their life represents the dual burdens of women in the community, constituting the truth of their social reality. The production and representation of knowledge about the activities and experience of these women demonstrate a contextual picture of the study setting. It is equivalent to the 'sciences from the below', as argued by Harding (2008: 15), which prioritizes knowledge production on women's experience at grass-roots level (e.g. the remote setting of a territory that is equivalent to the Shora). In interpreting Mohanty (2003), the standpoints of the *Jele-Baoalie* women, including their indigenous knowledge of collecting *pona*, depict the knowledge production on the fishing community women of the less developed countries.

3.3.2 Divorced Women

The in-depth interview data reveals that men wed more than two times and thus turn polygamy into a common practice in the Muslim community in the Shora. Forest-dependent men prefer to wed teenage girls to gratify their sexual pleasure and believe that fish collected from the river, enriched with iodine and protein, increase the requisite sexual 'power' for intercourse. The bodily attraction between men and women or teenage girls frequently leads to unprotected sexual relations, resulting in unwanted pregnancies. According to the local customs, men are obliged to marry pregnant girls or women by divorcing their previous wives or by their wives granting permission. Divorced women never receive financial spousal support from their husbands, and after divorce a woman must return to the father's house, unwelcome and without alimony. This defines them as a burden and results in social exclusion and stigmatization by the men and the married women in the locality.

Bidhoba Polli[1] offers divorced women the opportunity to live together in a supportive and friendly environment. A group consisting of four to five divorced women rent a boat for 200 Bangladeshi Taka[2] per day to go to the forest. This group of women calls the Sundarbans forest *jongol* (jungle), and they compare the jungle to their best friend, as it provides them with resources and the best option to generate income. Analysis of the observation data reveals that the divorced women have believed forest-connected myths for many years, as they worship the goddess *Mabonbibi*, a part of Hindu religion that believes in the presence of the almighty goddess in the forest. This is significant to divorced women, as *Mabonbibi* is a symbol of women's power in the forest. Before entering the forest, the divorced women call *Mabonbibi* by making their obeisance and by the silent recitation from the holy Quran. If they pray before sunrise and sacrifice a live hen or goat to the forest in the name of *Mabonbibi*, this safeguards them against the dangers of the wild animals in the forest. Three or four boats of women enter the forest simultaneously in a synergy that is a good way to work collectively in the forest.

The groups of divorced women go as far as a quarter of a kilometre inside the forest to collect olives. They also collect *kawra* (*Sonneratia apetela*) fruits, *omora* fruits, *goran* (*Ceriops decandra*) timbers and small branches of the mangrove trees. They carry sacks and small bamboo baskets for collecting the resources to take back home. Salted green olives are a tasty popular food eaten by women and girls in Shora in the afternoon, whilst they cook ripe olives to eat with boiled rice. Sometimes they sell both green and ripe olives in the market to earn money. *Kawra* is a sour kind of fruit used in pickle making by divorced women. At first, the green *kawra* must dry in the sun so the wet layer can peel off, and then chilli powder and soybeans are added to make it spicy. The women sell the pickle they produce in the local market of the *Guburu Union* as another form of income. As divorced women do not possess furniture-making skills, they sell the wood of the *goran* tree at a cheap rate to the local agents of the timber houses, and use small branches of broken mangrove trees as firewood for cooking rice at the shelter home.

The male-dominated society in the studied region makes the social acceptance of the divorced women living at the *Bidhoba Polli* precarious. As patriarchal norms and values place men in a breadwinner position both in the family and in the community, the conservative Islamic outlook of the community members, and the lack of inclusion of women in the local governance system (i.e. *Union Parisod*), means the divorced women must arrange their own agency and social skills in the community. Therefore, the institutions ignore what the women do and how they contribute to society. These findings link to Harding's (2004) assumptions,

---

[1]This is a refuge constructed by NGOs to provide a communal home for the divorced women.

[2]The Taka (TK) is the official currency of Bangladesh.

emphasizing the constitution of the female groups' perspectives, and the power of dominant social and political institutions (e.g. patriarchy, local governance). They clearly demonstrate the social exclusion and marginalized position of the divorced women in the study setting where, as an oppressed group, they lack recognition from the community or the state.

3.3.3 Married Women

The participants report that with little schooling and information about familial responsibilities, most girls under the age of 20 marry in the region. In the post-marital phase in Shora, wives take over conventional family roles, including cooking for the family members, giving birth, caring for children and nurturing in-laws in the household. It is apparent from the observation data that in the post-cyclone landscape, the participants in the visited households struggle to achieve three meals per day. The hardships caused by either the extreme poverty, or the natural disasters, has meant wives working inside the households have been forced to engage in paid and unpaid labour outside the home, even though they do not go to the forest. Such unpaid tasks include cooking for the household members, fetching water from a distant deep well, producing vegetables through home gardening and bearing the responsibility for their children's schooling. Rural women have to execute their tasks to hold the familial bonds together to a standard which meets with their husband's approval, so, in addition to their household chores, they regularly spend more than 4 to 5 hours a day on outdoor tasks, making it difficult to find time for paid work. This practice of doing unpaid jobs outside the home started after Sidr extirpated crops, vegetation and other sources of living, degrading the already limited local resources in the area.

The sources of paid work that women have include crab catching, goat rearing, sewing mattresses, making clothing, including saris and shayas, and owning grocery shops. Generally, women from an economically comfortable background are unlikely to be involved in such activities, but women from poorer families are encouraged to take on paid work. The limited number of women interested in earning money by their own efforts take a small loan[3] from microfinance programmes offered by several national and local NGOs, including Grameen Bank, Broite, Asha and Muslim Aid, to enable them to operate income-generating activities.

Participant: 'The daily earning of my husband (a boatman) is too low for us to afford the cost of food for the five members of our household. Earning barely enough money even in the summertime when the tourists from the town come to see the forest, he always finds it difficult to lead a good life. I recall starving, along

---

[3]In the location being studied, a small amount of cash, such as 5,000–10,000 TK, is given to women to operate businesses or income-generating activities.

with my children, for the want of three square meals a day. I could not tolerate the suffering of the children. Taking a loan from Grameen Bank, I started a small grocery shop where Shora's women and men come to purchase daily essentials (rice, spices, ropes, muri, chira, sugar, and salt). Now I earn minimum of 150–200 TK a day. It has enabled me to provide an ancillary economic support to the family'. (A female participant, age 55, owner of a grocery shop.)

3.3.4 Men's Forest-Going Practice

The Brotie[4] household survey (2011) demonstrates that 40% of men from 1,380 households constitute the entire population of Shora, showing that men head most of the households. When asked about their understanding of the forest, the male participants pay specific attention, as although the marauding Royal Bengal Tigers lie in wait, it is a very beautiful forest for them to roam through and from which to earn an income. The male participants articulate that the green leaves of plants appear stunning on a sunny afternoon, and viewing the loping movement of the *mayibiehorin* (deer) in the daylight is an extraordinary spectacle.

The male participants like to discuss when and how they visit the forest and what they do during their time in the forest. From the in-depth interviews, it emerged that the men go to the forest for periods of less than 2 days or periods longer than 7 days in the dense part of the forest.

Regarding the short periods, the participants note that during summer months, especially in May, June and July, they have to wake up before 3 am. After saying *Fazar* prayer (dawn Islamic prayer) and eating breakfast with *pantavat* (fermented rice), the men prepare to leave for the forest. The participants use *lungi* (sarongs), napkins and a fitted shirt as their uniform. They also anoint their entire body with mustard oil to keep it slippery, a process that they believe safeguards their skin from the attack of poisonous insects in the forest water. A group of 10–15 married men over 35, and some younger men aged around 25–30, hire a motorboat for ceremoniously entering the forest.

Before stepping on the motorboat, they take entry cards issued by the local forest office, axes, ropes, shovels, knives, baskets, nets, handmade traps with *borse* (a locally manufactured fishing line), pieces of beef to be used as food for crabs and drinking water. Initially, they drive the motorboat very fast to vie with the sun rays to reach the canal. The moment they get closer to the bank of the canal, everyone recites the *Aitalkurse*[5] or says the *dowa* of the prophet *Younus*. They alight from the boat with their equipment and begin felling the timbers of the *hetal (Phoenix paludosa), kawra,* and *sundori (Heritiera fomes)* trees. It takes approximately 3 to 4 hours to collect timber from its main roots, and after felling the tree, men strike at

---

[4]Brotie is an NGO working with the cyclone victim women and children in the village Shora.

[5]*Aitalkurse* (or *Ayatul Kursi*) is an Islamic verse written in the Quran.

the upper part of the trees with their sharpened axes. This process also involves shovels to separate the timber from soil and thus requires adequate muscular strength. Afterwards, the men strongly knot the separated timbers with ropes for placing it under the motorboat.

Once the high tides from the river replenish the canals, the men depart with the timbers they have accumulated. While returning, each group forks into two sections on separate boats for placing the handmade traps in the river to hunt crabs. The men add a piece of beef to the *borse* for baiting the crabs into the traps. Whilst this is legal behaviour, illegal logging occurs deeper in the forest, and the poor honey-collector community, *Mawali*, breach the law by going into the forest during the *Bengali* months of *Baishak* and *Chattro*, as these are the best months for gleaning honey.

Participant: 'I have seen my Mawali (honey-collector) father gathering honey from the deepest part of the forest. Collecting honey is a challenging task, as bees in the hive bite the collectors. However, before going to the forest, my father, along with his friends, planned a stay in the suitable corner of forest where the hives of honey were likely to appear. After the planning phase, they procured the foodstuffs and requisite cooking and sleeping instruments and loaded them on the motorboat. They set off for the journey at midnight, when the river, Kholpatura, swelled up with high water. It took them 48 hours to reach their destination. Before harvesting honey, the team uttered the *Aitalkurse* as a talisman to protect the area from the sudden assault of marauding animals. They collected honey by using their indigenous skills. My father and his mates once got home from the forest more than a week later. I found one of my father's legs missing. Agitated, I stammeringly asked him what had befallen him during his sojourn in the forest. Recounting the gory details of the incident, he replied a marauding tiger mutilated his leg. 'Fortunately', he gasped, "it spared my life". (A male participant, age 75.)

3.4   Forest Serves the Basic Needs of Life

In 2013, the World Wide Fund for Nature (WWF) reported that forests cover 31% of the total land mass, provide livelihoods to 1.6 billion people and housing for more than 300 million people worldwide. The Sundarbans contains three kinds of important life-saving forest resources: honey; shrimp, crabs and numerous fish like *chati, renu pona, powa, med, tangra, balae, pangas, vetke, passea, vangan* and trees like *sundori, gaya, gain, dundol, posur, hetal, kawara, golpata, goran* and olive.

To the participants, honey is the most valuable gift of the forest, created by Allah and tasting like ambrosia. During a prolonged drought that parched the local tube wells, the participants suffered from a dearth of safe drinking water. Given that the groundwater was contaminated with arsenic and the water layer plummeted, the local population had to drink honey as a substitute for water. The female participants, who are often responsible for obtaining the family's water supply, opine that drinking honey saves them travelling long distances to fetch water. The participants drink the honey with bread for breakfast, and they believe the honey has herbal medicinal properties that cure 108 kinds of diseases that may afflict the human body.

The study setting is known as the 'White Gold' of Bangladesh because of the glut of available shrimp and women's professional involvement in the sector, and hence the observation data documents women's shrimp catching in the canals in the forest.

Participant: 'Shrimp-catching is one of the main sources of occupation for women in Shora. Small shrimps known as Chati and Renu Pona (fingerlings) are found in the saline water of the small rivers of the forest. They are caught twice a day. When the high water swells up with the canal water, we set up nets in order to catch the fingerlings. After we collect and bring them home, the local middlemen, locally better known as the Furi, purchase them from us. The middlemen then sell these shrimps to the local market at a far higher price than they buy them from us. Because women are not permitted to go to the market and are unaware of the current rate of shrimps, we are often duped by the dishonest middlemen. With the pittance earned from catching shrimps, we have to eke out a living and afford education for our children at school'. (A female participant and shrimp-catcher, age 43.)

Both male and female participants note that the *powa, med, tangra, balae, pangas, vetke, passea, vangan* and *crabs* are traditional food sold at the *Dumuria Hat* (market), which meets every Saturday. These sea fish are quite popular because of their high levels of iodine, and as doctors prescribe anaemia patients iodine-rich food, catching fish from the small canals and crabs prior to 2009 would bring in high levels of income for both men and women due to high market prices and attractiveness to the European markets. Before Aila devastated the region, a female participant described the crab-catching community is the richest class in the study area.

Participants believe that the exquisite beauty of the *sundori* trees attracts tourists from around the world, who enjoy a boat trip into the forest to savour the extraordinary spectacle of the mangrove plants. Ferrying tourists across the forest provide a good income during the summer months, and during the boat trip in the forest, tourists love to see golden deer, *modantak* birds and owls. They also enjoy seeing monkeys cavorting from one tree to another and the activities of the journeyman *golpata* collectors. After being on the boat for 2 to 3 hours, tourists often end up visiting the Nildumur Camp, a menagerie of tamed monkeys, forest birds and crocodiles.

The Sundarbans are believed by many to be named after the prevalent *Sundori* tree, and the timber of the *sundori, gaya, bain, posur* and *hetal* trees is very useful because it is resistant to moisture, and hence aristocratic urban people buy it at higher market prices to construct their urban houses. The prospect of earning a higher price motivates the local people to fell the *sundori* trees for commercial purposes whilst they collect *gaya, bain, posur* and *hetal* trees for firewood to cook food at home and for making items like sofas, tables, cots and electric pillars to be sold at the market. It is important to note that wood for the pencils used by the schoolchildren is prepared from the *Dundol* tree, and the study participants repeatedly claimed that the raw material for the pencil-making industry in the divisional town comes solely from the forest. In addition, pickle as well as *tok*[6] comes from olive and *kawra* trees, with *tok* considered the best source of vitamin C in Bangladesh.

3.5 Illegal Activities Inside Sundarbans

As poorer people of the local population in forest areas are most likely the main users of forest resources (Munanura et al. 2018), the constraints on their livelihood influence them to engage in illegal forest use. This is similar to the situation in other developing countries (Munanura et al. 2018) and includes illegal felling of trees and bribing forestry officials. Illegal felling of trees results in gradual deforestation, and eventually the extinction of local wildlife. Whilst poorer people may actively engage in felling, the behind-the-scenes cartels perpetuate the economics and financial transactions (Islam/Sato 2012). Although the Forest Act (1927) enacted during the British colonial period and the 1974 and 1990 amendments were designed for the preservation and development of the Sundarbans, political instability, weak administrative forest policies, self-perpetuating grand corruption and the bureaucratic malfeasance and their collusion with the cartels has resulted in ineffective legislation and recommendations yet to be implemented (Ahamed/Ahamed n.d).

Despite the environmental impact, infraction of forest law by poorer people and law enforcement officials is common practice in Bangladesh. The in-depth interviews revealed that approximately 95% of the local population and forestry officials violate the law concerning forest resources for their own financial gain. The participants claim that the forest department pays the forest officials a meagre salary, so they supplement their income by selling valuable timber to the dishonest businesspersons in the area. Moreover, participants claim that officials take bribes for endorsing the requisite forest entrance letter, and this continues throughout the year. The local police and leaders support this, and hence the officials share the bribery money with them, knowing they will not face the legal system. The focus group data related how forestry officers insist on disallowing women into the forest, as they do not pay the bribe for permission letter approval. Consequently, the officials do not protect the women's boats from marauding animals but they do protect the men's boats through the security of the bribe. Furthermore, when on-duty forestry officials locate any non-permitted women's boat in the forest, they capture it to extort money from the women who want the boat released back to them in one piece. Despite the stipulations of the forest department or the NGOs working in the

---

[6]*Tok* is a soured liquid eaten with rice at lunch in the rural villages of Bangladesh.

region to use the Sundarbans in a sustainable way, there is no apparent systemic training to achieve this. As such, the lack of knowledge on the scientific use of the Sundarbans forest leads the participants to continue with the traditional practices for gleaning forest resources, inadvertently damaging the forest.

3.6   Customary Use of the Mangrove Forest Resources

The Forest Peoples Programme (2013) brought into focus the indigenous communities' traditional knowledge, customary practices and use of natural resources. As the specific types of knowledge, and practices of rituals and beliefs, are usually unwritten traditional rules and regulations, this study attempted to document the local population's traditional practices regarding the use of the Sundarbans forest, as it contributes a large portion of their livelihoods. During observations of men cutting wood, they used sharp axes to separate wood from the big trees inside the forest, with the axe handle made from the branches of the *amur tree*, found in the Sundarbans. Sometimes they use a *da* (similar to an axe, but smaller) not only to cut wood, but also to split it for firewood. Men who cut wood also use the *dinghy* to shift the wood from the inside of the forest through small canals towards the main boat.

The participants consider *golpata* (*Nipa palm*) the most versatile and valuable non-timber tree in the Sundarbans, because it provides excellent thatching material for cottage walls and roofs (Kabir/Hossain 2008). *Golpata* usually grows in tidal channels, rivers, low-saline bays, marshy interiors and moderate saline in freshwater zones of the Sundarbans. The village houses roofed by *golpata* are not only in the forest region adjoining the Sundarbans but also in the Barisal Division. The edible *golpata* fruits ripen during July and August, and are culled to prepare herbal medicines for the treatment of indigestion, constipation and other intestinal disorders.

According to the in-depth interviews, *golpata* collection occurs mid-November to the end of March, but the Forest Department permits 45 days to collect *golpata* from demarcated portions of the forest. If the *golpata* collectors flout the order of the authority or fail to observe it, they have to pay the forest office 300 TK (Bangladeshi currency) for each day. Each letter of permission allows a group of 15 men to collect the *golpata* and, once granted permission, some 25–100 boats led by a *sardar* (head boatman) form a fleet to venture into the forest for 30–50 days.

Before commencing the journey, each *golpata* collector eats a meal of boiled rice and milk with molasses prepared from date juice. After the meal, they pray at the local mosque, as they believe prayer will protect them from the attack of wild animals. The team takes necessary food, kerosene stoves for cooking, medicine and potable water. Their equipment includes *gasidas* (very sharp iron machete-like weapons), axes and ropes. At sunrise, the boats enter the forest, and, once cutting starts, they only cut leaves longer than 9 feet long. Each *kahon* (local measuring unit) contains 80 *pata* (divided leaves) and a team regularly collects 30 *kahons*, so when all the boats are replete with thousands of *kahons* within the allocated time, the fleet departs the forest. They anchor the fleet near the *golpatapotte* (local market) and leave the *golpata* in the sun to desiccate so that they can sell it to local buyers at a good price.

3.7   Changes in the Customary Use of Forest Resources

Agriculture contributes 19.6% to the national GDP of Bangladesh and provides employment for 63% of the population, but it is heavily reliant on the weather. Harvests are at the mercy of the cyclones that hit the country's rural villages.[7] Therefore, the villagers living in extreme poverty in the coastal belt have to prepare for and fight against the natural disasters to maintain their sources of livelihood. *The Daily Pattrodut*, a popular local Bengali newspaper of the Satkhira district, reveals that Shora is the village most susceptible to cyclones in the country, and hence the researcher follows the victims' narratives to document the changes occurring in the use of forest resources.

Before the cyclones in 2007 and 2009, there was ecological harmony between the forest and the local ecosystem of Shora. The participants recalled that, although the summer temperatures were swelteringly hot, the steady blow of cool breezes from the riverside and the shade of the trees provided sufficient cooling. The participants also recollected that on the way back home from the fields they used to stop under the trees to ward off the exhausting monotony. During the rainy seasons, the precipitation was suitable for the crops to thrive and the participants remembered going to the forest to cut the branches of trees to make essential agricultural equipment, such as spades, harrows and *kasi* (scythes). When harvesting time arrived, the wives accompanied their husbands to collect rice from the paddy fields, amassing adequate income, leading to a comfortable life.

Participant: 'I was once a very lucky farmer because the crops yielded optimum production in the locality. Although it was very difficult to invest the requisite amount of money to produce crops twice a year, we had enough precipitation to irrigate the land, which cut the expenditure drastically, making all this possible. Since I had little time while working in the paddy field, my wife brought food for me to dine outside. While returning home, I used to sit beneath the banyans or date trees to cool off and let the sweat on my body evaporate. The shade of the trees served as a natural air-cooler. Overall, before 2006, most of the farmers like me at Shora were very happy'. (A male participant, age 35.)

Before the cyclones, the participants were largely dependent on the agrarian industry for sustenance. The arable lands were fertile for sowing paddy, wheat, and local vegetables, such as *puishak, lalshak, datashak, kosu, lau, kumra*, and *borbote*.

---

[7]Source: http://www.nationsencyclopedia.com/economies/Asia-and-the-Pacific/Bangladesh-AGRICULTURE.html#ixzz2OxMY1bxm.

The participants claimed that the yields from mangos, papayas, coconuts, palm trees and bananas were enough for the community to mitigate possible vitamin deficiencies.

Participant: 'In comparison to most other inhabitants of Shora, I had more dairy cows in the cowshed. My pond was more abundant with fish (e.g. salmon, rouhe, mrigel, and tilapia); my land was replete with paddy and I had ready-cash in hand; rice from the field and vegetables from the homestead would make up our satisfying everyday meal. In addition, there was the smooth flow of cold air in the village and the shade of the trees made life happy and wealthy. Moreover, women were engaged in seasonal jobs (e.g. digging soil, serving as home-assistants, Food for Work programme of CARE Bangladesh etc.). Only a few men and women with no land of their own or other sources of income used to go into the forest'. (A female participant, age 33.)

Before the cyclones, the participants depended less on the forest for their livelihood, since the crops from the field and other paid jobs were viable alternatives, and hence accessibility to the available resources restricted the local population from using the forest for greater scale sustenance.

The 2012 Special Report of the Inter-governmental Panel on Climate Change (IPCC) discovered the social as well as physical dimensions of vulnerability resulting from the weather- and climate-related disasters, with the report claiming that the extreme weather and climatic events influence the increasing threats to populations and assets. Similarly, the participants in the study location had to endure inhumane conditions in the aftermath of cyclones Sidr and Aila. Floodwater inundated cultivable lands for long periods and, due to higher saline levels in the water, this caused a loss of land fertility, trees and houses. With the crops and paddy fields devastated, there was a surge in the price of everyday essentials, and, with a dearth in income, countless families had to move away from the district to find work. In some instances, when the male family members left to find work, the wives stayed at home, eking out a miserable livelihood on pittance wages. One female participant alleged her husband married a girl in the city so he could live a comfortable life there, away from the poverty inflicted by the catastrophes. Other wives in Shora experienced similar circumstances, through either divorce or becoming widowed, meaning they faced social exclusion. To create a livelihood, these women had to go to the forest, traditionally a man's job, and although they initially began collecting shrimp for their income, they expanded into collecting leaves, fruits and small timbers. The participants claim they now feel a sudden increase in the temperature resulting from natural disasters, possibly due to the lack of trees, which is turning the area into a desert-like existence. The participants liken the connection between nature and the local population to an umbilical cord that has now snapped. This has created an imbalance, and subsequently, the water is unsafe to drink. Since the traditional occupations no longer exist in the region, the local population has had to find a different source of income or venture further into the forest to collect resources.

Participant: 'The sudden attack of pirates and subsequent demand for a ransom inside the Sundarbans is a common phenomenon that puts a frightening challenge before the inhabitants of the locality. Before the Sidr, people, especially the women, never went to the forest. It is very daunting for a woman to be in the forest among the lurking pirates, but the calamity-induced poverty has left almost no alternatives but to go to the forest'. (A female participant, age 39.)

3.8 Conclusion

In conclusion, the senior members of the society acquaint the younger generation with the forest-going myths by passing down the myths from generation to generation. Although it is evident that men go to the forest, the *Jele-Baoalie* and married women also now venture into the forest, but in smaller numbers. Some venal officials acting in collusion with administration and law-enforcing agencies has led to corruption surrounding the forest-going tradition. Regarding the customary use of the forest, natural disasters in recent times have compelled a greater number of people to go to the forest. However, their ventures often end in fiasco when pirates like the venal forest officials try to impede their activities. In the post-disaster landscape, the displacement of the men from the study area to the district town in search of a higher income may have influenced the increasing number of divorced women, despite the deeply entrenched patriarchy marginalizing divorced women, who continue living on the periphery of the society.

References

Ahamed, U.M. and Ahamed, F. (n.d.), *Control of Illegal Activities in the Sundarbans Forest of Bangladesh: Overview of the Regulatory Aspects*, Seminar Paper, Department of URP, Khulna University, Khulna, Bangladesh.

Forest Peoples Programme (2013), *Supporting Forest Peoples' Right*, Indonesia, viewed 17 Apr. 19. http://www.forestpeoples.org/en/resources?Publications[0]=work_theme_and_topics_pu blication%3A407&Publications[1]=language%3Aen.

Harding, S. (2008), *Sciences from Below: Feminisms, Postcolonialities, and Modernities*, Durham, NC: Duke University Press.

Harding, S. (2004), *The Feminist Standpoint Reader* (ed.), New York and London: Routledge.

Islam, K.K. and Sato, N. (2012), 'Deforestation, Land Conversion and Illegal Logging in Bangladesh: The Case of the Sal (*Shorea robusta*) forests', *iForest*, Vol. 5, pp. 171–178.

Kabir, H.M.D. and Hossain, J. (2008), *Resuscitating the Sundarbans, Customary use of Biodiversity and Traditional Practices in Bangladesh*, Dhaka: Unnayan Onneshan; The Innovators.

Mohanty, T.C. (2003), *Feminism without Borders*, Durham, NC: Duke University Press.

Munanura, I.E.; Backman, K.F.; Hallo, J.C.; Powell, R.B.; Sabuhoro, E. (2018), 'Understanding the Relationship Between Livelihood Constraints of Poor Forest-adjacent Residents, and Illegal Forest Use, at Volcanoes National Park, Rwanda', *Conservation and Society*, Vol. 16, No. 3, pp. 291–304.

Sultana, M.A. (2010), 'Patriarchy and Women's Gender Ideology: A Socio-Cultural Perspective', *Journal of Social Sciences*, Vol. 6, No. 1, pp. 123–126.

Chapter 4
Women's Use of the Sundarbans Forest Resources

Abstract This section of the book is a detail description of the whole process of the women's resource collection from and use of Sundarbans forest. The women of Shora community have different layers of activities related to Sundarbans forest, e.g. preparation, resource collection, resource processing, resource selling, household use of the resources and so forth. Women of Shora community have to go through a continuous struggle with men in the market place and society to make their living. Women play a great role in the conservation of Sundarbans forest as the empirical data shows that their activities are not harmful to the forest ecology. To conclude, women of Shora community are overcoming the established gendered norms through their unique financial activities on Sundarbans forest, and their developed knowledge is helping them in this whole process.

Keywords Forest resource · Gendered norms · Forest conservation · Sundarbans forest

4.1 Introduction

The previous chapter has clearly pointed out the informants' indigenous perceptions of the Sundarbans forest, and described the use of forest resources. This chapter takes into account the empirical data connected with women's interactions with the forest, and attempts to document women's behaviour regarding their use of the forest resources in the forest and at home. Afterwards, it presents how informants' decision-making attitudes are suppressed in the home and the marketplace, and how they conserve the forest for future generations. In addition, it aims to interpret the findings based on feminist political ecology theory, and integrates relevant archival literature among sections of the chapter to support the arguments for revealing the contextual (study setting) norms and values of the informants.

S. Roy, *Climate Change Impacts on Gender Relations in Bangladesh*,
SpringerBriefs in Environment, Security, Development and Peace 29,
https://doi.org/10.1007/978-981-13-6776-2_4

4.2 Primary Resource User

This section focuses on the activities of female forest users at home and in the Sundarbans forest, placing them as the primary users of the mangrove forest resources. The frequent use of firewood, logs, timbers, as well as dried leaves, by women in the forest areas indicates that they develop adequate knowledge of the forest resources (Wan et al. 2011). Previous studies (Edmond 2008; Godfrey et al. 2010, quoted in Wan et al. 2011; Gbadegesin 1996; Aluko 2018; Tyagi/Das 2018) on how the women use the forest have neglected the indigenous knowledge they possess and apply in preparing forest resources for the consumers of the market. In the same way, the existing body of literature has overlooked women's interactional perspectives on the mangrove forest resources in the context of Bangladesh.

According to Aluko (2018), indigenous women are considered a valuable part of society. Women's indigenous knowledge has been used since primitive times throughout the world for survival activities. Indigenous knowledge is used in various fields, including agriculture, small industry, health care and family maintenance (Aluko 2018). Gibb (2007) emphasizes that as women are more involved in non-commercial and household activities than men, they possess more specialized knowledge of natural resources. According to Myers (2002), environmental security is pivotal to the blanket security of any country. Nnadi et al. (2013) suggest that there is a close connection between indigenous knowledge, environmental security and sustainable development. Proper use of indigenous knowledge will result in successful sustainable development, according to research conducted in Nigeria. The indigenous knowledge of women may bring positive outcomes in achieving sustainable development goals. Olatokun/Ayanbode (2009) discovered that many of the development policies of Nigeria failed because they did not consider indigenous knowledge carefully. Implementation of indigenous knowledge also showed positive results in Tanzania (Asogwa et al. 2017). Indigenous people possess a belief that all living objects on earth are holistically connected. According to the Food and Agriculture Organization of the United Nations (2009), indigenous agriculture, farming and forestry are backed by long-established norms and practices. Indigenous knowledge plays a crucial role in preserving agricultural diversity, natural resources and livelihoods.

Aluko (2018) holds that the traditional medicinal systems of many communities are based on indigenous knowledge. Eyong (2007) notes that most of the global population depends on an indigenous medicinal system. Indigenous knowledge and norms also help in water conservation. Lactating mothers keep away from water bodies, and other measures taken by indigenous people include not washing near rivers or streams, and not throwing dirt into water (Cheserek 2005). An African-based study found that indigenous knowledge of agriculture delivers better results compared to 'modern knowledge'. Modern knowledge and technology can only work when they take into account indigenous knowledge in a respectful way (Olukoya 2006). Many international development programmes have included women's knowledge in their plan (UNDRIP 2007). Women can contribute in developing new knowledge and techniques (FAO 2005). Escobar (1995) suggests that all development programmes should examine the local context and consider indigenous knowledge. Ibnouf (2008) argues that environmental and development challenges can be confronted with women's gender-specific indigenous knowledge and skills. Denton (2002) emphasizes that to achieve sustainable development goals, women should be included and their indigenous knowledge should be valued in development programmes. Aluko (2018) further argues that policymakers should consider indigenous knowledge and practices to ensure fruitful planning. Women's indigenous and traditional knowledge should be used to promote and protect biodiversity. To achieve this, women should be empowered, and there should not be any gender discrimination (Aluko 2018).

Upadhyay (2005) argues that the rural women of South Asian countries possess considerable knowledge of the characteristics, distribution and site requirements of indigenous trees, shrubs and herbs. Their conventional understanding of the use of plants for food, fuel, medicine and crafts plays a leading role in conserving the variety of species in accordance with their usefulness to their community. Further, they understand which forest products to use for familial diets, and during natural calamities, such as famine, flood and cyclones. The foods women preserve not only serve the familial needs but are also sold to the market for cash income (SD and FAO 2013). Beyond their everyday domestic chores, women invest their labour in preserving forest goods in preparation for critical circumstances. All this illustrates that the women in forest areas are not only involved in performing the multiple tasks associated with processing the forest products used for familial needs, but also act as the agents of those selling those products at market. Similarly, the female informants in the study location can be considered primary processors of the collected resources (fruits and firewood). The observation data reveals that three to four women constitute a group and collect three to four sacks of green olives from the forest at a time. Each sack is filled with olives weighing between 30 and 40 kg. These are carried back home from the forest by the female informants. As soon as the olives are poured out of the sacks, they are equally distributed between the group members. Thereafter, to make tasty food, each woman uses water to separate the olives from mud before exposing them to the hot sun to dry (Fig. 4.1).

The forest-dependent women claim they perform the drying process without their husbands' intervention. After olives are dried, mustard oil, chilli powder and sugar are blended with the olives to create jam, jelly or pickle to augment the household diet. The informants' husbands sometimes sell the concoction in the local market. The entire process of making jam, jelly or pickle makes it evident that the informants work more than 3 hours a day, 6 days a week to produce it. The firewood – often *Gaya*, *Posur* and *Baine* – gathered from the forest is broken into small pieces by women for easy carrying back to their home. A sharp axe is used to break down the wood. Hitting the wood effectively requires muscular power, patience and skill, according to the informants. '*The fresh raw-wood we bring from the forest is too strong to split into tiny pieces. So, by axe, we strike Arkope at an angle to make it smaller than the original piece.*' (Fig. 4.2).

Fig. 4.1 Processing fruit. *Source* Researcher; copyright: Sajal Roy

The observation data confirms that the small pieces of firewood are kept in the scorching sunlight for a couple of days in the informants' courtyard. This dried wood is known as *Lakre* (firewood) by the area habitants. It is either used for cooking food for the household, or retailed to the men wood traders at a cheap price. Generally, during the rainy season – commonly from May to September in the country – the dried wood has a good market value. The high value of firewood is due to its scarcity as fuel for cooking and the production of goods for the industries in the district town. According to the informants, at the time of vending *Lakre* to the representatives of the city timber houses, they set up a wooden balance

Fig. 4.2 Processing firewood. *Source* Researcher; copyright: Sajal Roy

in the courtyard to measure the amount of wood to be sold. Afterwards, it is transported to the district town by a four-wheeled vehicle. Market value is measured according to 40 kg '*Akmone*' of *Lakre*, which is typically sold for between 45 and 60 TK. The money earned is possessed by the informants' husbands; it follows that women's intensive labour in the processing of wood is devalued since they are deprived of the reward they are supposed to receive from the market (Fig. 4.3).

The *pona* caught by women from the forest's surrounding channels and rivers are kept in water in an aluminium pot to keep them alive. When the *Bagdha Pownas* are brought to Shora, they are handed over to an internal *furi* (middleman), the local agent of the fishing house located in the city centre, in exchange for a small amount of money. Prior to selling *pona* to the *furi*, the women use a snail

Fig. 4.3 Women's participation in *pona* processing. *Source* Researcher; copyright: Sajal Roy

Jhinuk to count the *pona* from the pot. In this activity, a special soap, Datolsaban or liquid Datol, is used to wash hands to keep the *Pona* clean and virus free.

The observation data reveals that 100 pieces of *Bagdha Pona* are peddled at 250–300 TK to the local *furi*. The local *furi* goes to the local market at Munshegonj and sells the same amount of *Pona* to the fishing depot for 500–800 TK, which ensures his profit. Interestingly, the *Pona* dealers benefit threefold by trading it to foreign buyers, according to the female *Pona* collector informants. It is widely recognized that the women involved in the preliminary processing of *pona* are indirectly cheated by the *furi* and *pona* dealers and used as moneymaking instruments of capitalism. Owing to the provocative attitudes of men towards the *pona*-collecting women, and the threat of humiliation outside the home, they dare not go to the fish depot in the city centre. As a result, they rely solely on the payment given by the internal *furi*. Consequently, those *pona*-catching women who overcome the obstacles at home and rivers become the recipients of the lowest market value. This scenario of gaining money by the ultra-poor *pona*-collecting women reproduces the *Marxist philosophy of proletariats*, as the surplus value obtained from the *Bagdha Pona* is not directly allocated to the real owners – the *pona*-collecting women – but instead shared by the internal *furi* at Shora and the fish dealers at the depot.

Participant: 'The pona dealers frequently receive a great deal of foreign currency by exporting our collected *Bagdha*. Being women, our presence at the fish market, which is commonly pre-empted by men, is not appreciated. We are to earn, and keep an acceptable social image in the locality. If we fail to maintain it, nobody will come to help us in an emergency. Therefore, instead of going to the marketplace, we cannot but agree to sell the collected *pona* to the internal *furi*. Although we are not getting the proper market value, there is nothing we can do about it.' (A pona collector woman, age 45.)

The above discussion elaborates on women's involvement in the processing of firewood and fish at home. The women informants, as primary resource users, apply their indigenous knowledge and skills in preparing the forest resources as products for market. Due to the social barriers mentioned above, women's labour is devalued and disrespected at home and in public. In addition, the knowledge they make use of for processing the *pona* and *Lakre* is followed by their junior fellow women over the years in the region. The transmission of knowledge about environmental resources from woman to woman is a trend in the location under study. Although the study attempts to reveal the women's knowledge about the processing of the forest resources, both in the forest and at home, it acknowledges men's in-depth perceptions on the collection and processing of timbers, *golpata* leaves, and honey from the deepest part of the forest, as mentioned in the previous chapter. Thus, the informants' behaviour towards the Sundarbans forest resources exposes a pattern of *gendered knowledge*, as coined by Rochelle et al. (1996).

4.3 Decision-Making Between the Market and Women Gatherers

The division of labour, in terms of gender, is a central feature of gender inequality (Cohen 2004). This explains how men become specialized in paid work within the market and women tend to be specialized in unpaid work within the home and marketplace (Washbrook 2007). The women informants were either involved in forest-connected activities for livelihood or performing domestic duties (as mentioned earlier) and are poorly remunerated for their physically taxing labour. Therefore, the division of labour and the meagre income from the household or outside has replicated an incorporeal trend over the years.

The forest timber products and non-forest products are mostly marketed for cash income. However, when gaining access to credit and its allocation to the forest user women and housewives, women's timid behaviour in front of men erodes their opportunities to obtain cash income at the locality. The weakness of women's attitudes and low expectation of reward for their invested labour have been indirectly abused by men over the years. Due to the low level of education, the lack of knowledge about the proper value judgement of life, and the fragile socio-economic status of the local area, women are always expected to support their husband's ideological position, which turns them silent in domestic affairs.

When attempting to visit a relative's house or consult a doctor in the city centre or a remote location, women seldom venture out of 'purdah', on their husband's recommendation. The informants report their reluctance to put on 'purdah', which makes it problematic to breathe pure oxygen, but they comply to satisfy their husbands' expectations, and show respect for Islamic ideology. The individual's clothing preference is ignored to satisfy their husband's norms and values. In this situation, the women informants have to navigate the existing local Islamic religious norms and values whenever they try to claim their actual financial dues from their male counterparts, including internal *furi*, husbands, fish dealers and agents. Due to this negotiated concern, men's powerful social position at Shora is deepened, which undermines women's sense of religious identity. This strengthens men's masculinity and simultaneously undermines the women informants' social position as mothers, caregivers and sisters. Individual woman is labelled by their husband, or another masculine representative of society, through regular practices (Butler 1990; McDowell 1995, cited in Radel 2012).

The in-depth interviews data connected to homemakers' domestic duties reveal that they are repeatedly treated as domestic servants by their husbands and in-laws. The men as the head *Kortabakte* of the household occupy the breadwinner's position and also become the key decision-makers and main role players in household matters. These include the behavioural patterns of wives, reproductive choices and options, when to be pregnant, and claiming the money earned from the forest resources processed by their wives, according to the informants. In the study site, these practices have turned into a deeply rooted social structure.

Participant: 'Before my marriage, my duties and responsibilities at home and outside were identified by my father. I had no freedom of expression of my own life. I had been guided to maintain the religious obligations imposed by the 'Huzur,' the head of local mosque and mentor of every family. I was frequently advised to follow the religious guidelines; nonetheless, it was my father who would have been summoned to the mosque if I had committed any misdeed. After my marriage, I was obliged to follow the rules to go to heaven after death; a wife is to stay under the regulation imposed by her husband and directed by her husband's house.' (A woman informant, age; 45.)

This quotation confirms that women are bound and forced to accept the religious opinions and guidance of the male representative of the local society of Shora. The forest user informants, or women at home, never dispute this ideological dominance imposed by father or husband; rather they submit so as to sustain their social image. It is linked to the *Hegemonic Masculinity* perspective guided by Connell (2009), emphasizing the power relationships between men and women as well as men and men, where dominance of some men over women and other men is seen as an outcome of institutionalized social structures, as reflected in the narrative of the above informant.

The decision-making of women in household-connected issues in rural regions of Bangladesh is contingent upon their financial contribution to the family. However, it appears from the study context that women are not welcome at their nearby or remote marketplace. This constrains women's ability to gain financial output from the market, as it robs them of the opportunity to cut out the middlemen and earn an equitable cash income by selling the product they process to the customer direct. Because they live under social circumstances that hardly allow them to be visible in the marketplace or earn a decent living, their voices are rarely heard in the familial decision-making process.

The women's ideological position in performing household activities, and instructions on how to treat male family members at Shora, are generally dictated by the husbands or fathers-in-law. Though this conservative outlook of patriarchy tethers women to the domestic sphere, the resources they collect for use in the household, and the environmental concerns they have, represent women's intimate connection to the forest ecology, and self-environmental care for their children. This depiction of feminine forest-related activities in the subsistence economy at Shora harmonizes an integrated system to gratify the basic needs of the people (Dankelman/Davision 1988; Shiva 1989, quoted in Nhanenge 2011).

4.4 Women as Natural Conservators of the Sundarbans

This section captures the informants' activities with regard to the conservation of Sundarbans forest. Maiden (2011) suggests that women adopt environmentally friendly practices, such as terracing and *taungya*, the cultivation of fodder trees and leading campaigns against tree grazing. The author's study, conducted in East Africa and South America, identifies women as the main most regular gleaners of the forest product. In addition, the study finds that women are the key role-players in restoring the degraded lands in the community forested areas.

The focus group data indicates that a couple of women organize the courtyard meeting (Uthan-Boithak) twice a month, at which they discuss the preservation process of the forest. The *Uthan-Boithak* consists of female forest users and non-forest-going women. Key issues include deer safety, oxygen supply to trees, the necessity of tigers for the Sundarbans' wildlife, the corruption of forest officials and the significance of the forest for children and the future generation. It is claimed that the women at Shora are alarmed by the consequences of the natural calamity and the deterioration of the forest.

Participant: 'It is rapturous to see the movement of a group of deer that adds to the beauty of the forest. There are no official statistics about the actual numbers of deer in the Sundarbans. The men ranging into the deepest portion of the forest are business-minded. Though the Forest Department prohibits the slaughter of deer, it is hunted by those business-minded men for vending the deer's meat and skin at a high price to the local elite and the city dwellers. We, the women, always motivate our husbands not to kill the deer as they are deeply connected to the forest ecology. The presence of deer helps to maintain the balance between biodiversity loss and the sustainability of the wildlife.' (A woman participant, age 40.)

There is no official record that the women have ever killed a deer, or used the forest resources for highly commercial purposes. They discourage husbands from killing deer inside the forest. It proves that the informants are environmentally aware and realize the necessity of the Sundarbans resources for the local community.

The focus group data affirms that the Royal Bengal tigers represent a distinctive character of the Sundarbans. Royal Bengal tigers are found only in the Sundarbans and are on the verge of extinction. Owing to unplanned deforestation by the forest area inhabitants, and gradual deterioration of the habitat in the Sundarbans, the number of tigers is decreasing daily, according to the informants. The women informants assert that some dishonest men at Shora kill tigers and traffic their cubs with the goal of earning a large amount of money. Although Bengal tigers are protected by laws guiding the conservation of the forest's wildlife, these laws are frequently violated by the area inhabitants and Forest Department employees. Furthermore, the informants add that during the summer each year tigers may cross the river to enter the village. The area inhabitants get frightened as soon as they come to realize the fact.

Participant: 'With Goran sticks, gun and the strong nets, the men's groups at Shora are extremely courageous. They keep themselves always prepared to attack any tiger entering the village. During summer nights, a man from each household patrols the village so that tigers might not enter the village.' (A female participant, age undisclosed.)

The I-PAC project, initiated by US-AID in 2004, deals with the integration of the indigenous people for tiger conservation in the coast belt of Bangladesh. It aims to train women to use local knowledge and expertise to protect the Sundarbans forest resources. This project's animators make a door-to-door visit in order to secure an appointment with a woman from each household at Shora. The project uses the participatory approach to encourage women to get involved in the sustainable maintenance of the Sundarbans forest. The trained women encourage their husbands to patrol the village at night, and suggest they refrain from attacking the tigers.

Participant: 'We advise the patrolling team members to create a lot of noise with their voices so that tigers will hear the sound and go back to the jungle.' (A woman informant, age 49.)

The continuous patrolling system has been working effectively in the region. As a result, tigers now rarely dare to cross the river, according to the informants. This approach demonstrates that women are indirectly involved in, and contribute to, the conservation of the most important resource of the Sundarbans – the Royal Bengal tigers. This reflects the women's attitude towards their environmental responsibilities (Rocheleau et al. 1996).

The discussed activities of *Uthan-Boithak* illustrate the efforts and environmental concern of women's groups to safeguard the Sundarbans and their members' self-motivation to work as a team. Furthermore, the *I-PAC* project that inspires women to motivate their husband to patrol the villages at night depicts women's leadership for the tigers' safety and wildlife preservation from grass-roots level.

The observation data explores how forest-going women plant mangrove seeds in the wetland to foster their germination and help them grow and thrive like the big trees in the Sundarbans. This plantation process is performed with great care, and the interaction between women and mangrove plants is compared to women's motivation to engage in environmental care. The informants believe that harvesting planted seeds reduce deforestation to a smaller scale and contributes to a greener environment for future generations of the forested villages. This activity, and the concern of women informants in the forest, underlines Mies/Shiva's (1993: 14) argument: 'Women have a deep and particular understanding of future generations and life on earth through their intimate connection between nature and experience.'

Participant: 'We are less energetic than a male forest-user. To gather timber and trees, a woman needs to be very robust to cope with the boating, resist being taking by the pirates and deal with the worse climate. We are very scared by the frequent movement of the pirates because they might kidnap us and ask our family's for a ransom. Therefore, we always get access to the nearby forest lands where mangrove leaves, fruits and fishes are available. Big trees and wild animals are the creatures of almighty God, and maintain the balance between the consequences of deforestation and ecological protection. Considering our role as mother and care-giver to the family, we do believe that planting mangrove seedlings enhances the balancing process, and it will keep the habitat and environment safe from the probable natural calamity.' (A woman informant, age 35.)

This informant identifies challenges that hinder women's access to the deepest part of the forest. These include the long distance between the residence and forest resources, the physical weakness and threats of violence from pirates. Clearly, women informants struggle to access the forest resources located in the distant part of the Sundarbans, despite claims that due to long-term poverty caused by *Aila*, a great deal of men in the study region has been involved in robbery. The difficulties produce a limited gender space for women informants in their quest for livelihood support and independence. The perspective of women informants' challenges is supported by Sultana's (2011) framework of feminist political ecology, which describes the messiness of everyday politics and struggles over the environmental resources of women.

4.5 Conclusion

It is evident from the findings that the women interact only with the forest resources that are easily accessible to them from the nearby forested area. Due to intimidation caused by social forces and the threat of stigmatization, women informants rarely gain access to the deepest part of the forest. The gathered resources are used in the familial diet and sold at market. However, the market value forest user women are supposed to obtain from the internal *furi* or the fish dealers is not given to them. Nevertheless, the women informants, with their local knowledge, are closely engaged in conserving the forest resources. Not surprisingly, very few of the women informants interact with the forest resources collected by men from the deepest part of the forest. It is evident that women's socio-economic identity at Shora is controlled by the patriarchal ideology, but compared to men informants, women are more motivated and environmentally aware of the importance of wildlife conservation and preserving the wilderness of the mangrove forest.

References

Aluko, Y.A. (2018), 'Women's Use of Indigenous Knowledge for Environmental Security and Sustainable Development in Southwest Nigeria', *The International Indigenous Policy Journal*, Vol. 9, No. 3. https://ir.lib.uwo.ca/iipj/vol9/iss3/2, https://doi.org/10.18584/iipj.2018.9.3.2.

Asogwa, I.S.; Okoye, J.I.; Oni, K. (2017), 'Promotion of Indigenous Food Preservation and Processing Knowledge and the Challenge of Food Security in Africa', *Journal of Food Security*, Vol. 5, No. 3, pp. 75–87.

Cheserek, G. (2005), 'Indigenous Knowledge in Water and Watershed Management: "Marakwet" Conservation Strategies and Techniques', in: G. Förch, R. Winnegge, and S. Thiemann (eds.), *DAAD Alumni Summer School 2005: Topics of Integrated Watershed Management*, pp. 25–33.

Cohen, P.N. (2004), 'The Gender Division of Labor: "Keeping House" and Occupational Segregation in the United States', *Gender and Society*, Vol. 18, No. 2, pp. 239–252.

Connell, R. (2009), *Gender: In World Perspective*, 2nd Edn., Cambridge: Polity Press.

Denton, F. (2002), 'Climate Change Vulnerability, Impacts, and adaptation: Why Does Gender Matter?', *Gender and Development*, Vol. 10, No. 2, pp. 10–20.

Dyubhele, N.; Le Roux, P.; Mears, R. (2009), 'Constraints to the Economic Activities of Women in Rural Areas: IKS Community Development and Resilience', *Indilinga: African Journal of Indigenous Knowledge Systems*, Vol. 8, No. 2, pp. 230–240.

Escobar, A. (1995), *Encountering Development: The Making and Unmaking of the Third World*, Princeton, NJ: Princeton University Press.

Eyong, C.T. (2007), 'Indigenous Knowledge and Sustainable Development in Africa: Case Study on Central Africa. Tribes and Tribals,' *Tribes and Tribals Special*, Vol. 1, pp. 121–139.

Food and Agriculture Organization of the United Nations (FAO) (2005), *Building on gender, agrobiodiversity and local knowledge*. Retrieved from http://www.fao.org/docrep/005/AC546E/ac546e08.htm.

Food and Agriculture Organization of the United Nations (FAO) (2009), *FAO and Traditional Knowledge: The Linkages with Sustainability, Food Security and Climate Change Impacts*. http://www.fao.org/3/a-i0841e.pdf.

Gbadegesin, A. (1996), 'Management of Forest Resources by Women: A Case Study from the Olokemeji Forest Reserve Area, Southwestern Nigeria', *Environmental Conservation*, Vol. 23, No. 2, pp. 115–119. https://doi.org/10.1017/s0376892900038492.

Gibb, H. (2007), 'Gender Dimensions of Intellectual Property and Traditional Medicinal Knowledge', viewed 17 April 2019. https://www.undp.org/content/dam/rbap/docs/Research%20&%20Publications/poverty/RBAP-PR-2007-Gender-IP-Traditional-Medicinal-Knowledge.pdf.

Ibnouf, F.O. (2008), *Role of Women in Providing and Improving Household Food Security in Rural Sudan*, Unpublished doctoral thesis), University of Wales Swansea, Swansea, UK.

Maiden, J. (2011), 'Are You a Hunter or Gatherer? Understanding How Men and Women Use the Forest', *Forest News*, a Blog by the Center for International Forestry Research, Indonesia, Viewed 13 April 2013. http://blog.cifor.org/3734/are-you-a-hunter-or-gatherer-understanding-how-men-and-women-use-the-forest/.

Mies, M. and Shiva, V. (1993), *Ecofeminism*, London and New Jersey: Zed Books.

Myers, G.A. (2002), 'Local Communities and the New Environmental Planning: A Case Study from Zanzibar Area', Vol. 34, No. 2, pp. 149–159. https://doi.org/10.1111/1475-4762.00067.

Nhanenge, J. (2011), 'An Ecofeminist Analysis', in: Nhanenge, J. (ed.), *Ecofeminism*, New York: New York University Press, pp. 360–361.

Nnadi, F.N.; Chikaire, J.; Ezudike, K.E. (2013), 'Assessment of Indigenous Knowledge Practices for Sustainable Agriculture and Food Security in Idemili South Local Government Area of Anambra State', *Journal of Resources Development and Management*, Vol. 1. http://www.iiste.org/Journals/index.php/JRDM/article/view/9290.

Olukoya, O. (2006), 'Traditional Farming and Indigenous Knowledge Systems in Africa: Perspectives from the Ikale-Yoruba Experience', *Indilinga: African Journal of Indigenous Knowledge Systems*, Vol. 5, No. 2, pp. 157–166.

Radel, C. (2012), 'Gendered Livelihoods and the Politics of Socio-environmental Identity: Women's Participation in Conservation Project in Calakmul, Mexico', *Gender, Place & Culture: A Journal of Feminist Geography*, Vol. 19, No. 1, pp. 61–82.

Rocheleau, D.; Thomas-Slayter, B.; Wangari, E. (1996), 'Gender and Environment: A Feminist Political Ecology Perspectives', in: D. Rocheleau, B. Thomas-Slayter, and E. Wangari (eds.), *Feminist Political Ecology: Global Issues and Local Experiences*, New York: Routledge, pp. 3–23.

Sustainable Development Department (SD), and Food and Agricultural Organization of the United Nations (FAO) 2013, *Gender and Development*, Viewed 17 April 2019. http://www.fao.org/3/a-i3205e.pdf.

Tyagi, N. and Das, S. (2018), 'Assessing Gender Responsiveness of Forest Policies in India', *Forest Policy and Economics*, Vol. 92, pp. 160–168. https://doi.org/10.1016/j.forpol.2018.05.004.

United Nations Declaration on the Rights of Indigenous Peoples (2007). http://www.un.org/esa/socdev/unpfii/documents/DRIPS_en.pdf.

Upadhyay, B. (2005), 'Women and Natural Resource Management: Illustrations from India and Nepal', *Natural Resources Forum*, Vol. 29, No. 3, pp. 224–232.

Wan, M.; Coller, C.J.P.; Powell, B. (2011), 'Forests, Women and Health: Opportunities and Challenges for Conservation', *International Forestry Review*, Vol. 13, No. 3, pp. 369–387.

Washbrook, E. (2007), 'Explaining the Gender Division of Labor: The Role of the Gender Wage Gap', *CMPO Working Paper: 07/174*, pp. 1–89, Submitted to the University of Bristol, UK.

Chapter 5
Human Security, Sundarbans and Survival at Shora

Abstract As the people of Shora have faced several natural disasters in previous years, this chapter presents their knowledge and perceptions of environmental security and survival in the village, through an unfolding discussion on economic security. The people of Shora have expressed their concern and opinions about conserving Sundarbans and perceive that the forest protects them during natural disasters. Therefore, they are concerned that if Sundarbans is damaged, their lives will become insecure; hence the discussion considers the women's role in protecting the forest through care and indigenous knowledge.

Keywords Human security · Shora · Sundarbans forest

5.1 Introduction

This chapter unpacks the participants' in-depth perception of human security, their insights about the importance of environmental security at Shora, and the role of women in protecting the resources of Sundarbans. In discussing the core elements of human security, this chapter describes the benefits of and threats to human security in the context of Shora, which segues into discussion about the participants' experience in the pre and post Sidr and Aila contexts. In addition to the severe short-term impacts of cyclones, the long-term impacts have caused vulnerability in the forest ecosystem and the surrounding areas. Due to environmental threats and uncertainty in the region, the need for human security is discussed.

5.2 Human Security

The data from the focus group revealed that the participants consider the experiences of Sidr and Aila when discussing human security in relation to Sundarbans but prefer to describe the human security issues of the study environment. The

© The Author(s), under exclusive license to Springer Nature Singapore Pte Ltd. 2019
S. Roy, *Climate Change Impacts on Gender Relations in Bangladesh*,
SpringerBriefs in Environment, Security, Development and Peace 29,
https://doi.org/10.1007/978-981-13-6776-2_5

narratives demonstrate that the region's ecology has become vulnerable due to the cyclones, and this perception of the ecological setting provides one perspective of human security, as it is an ambiguous concept and used by scholars in various ways (Paris 2001). Tadjbakhsh/Chenoy (2006) note that this concept is mostly used by activists, as it has little importance to academics and politicians. Although James (2014) proposes that 'human security' should be related to military security or state security **as part of** the human condition, the concept emerged as part of a **narrower** paradigm of human development cultivated at UNDP, contributed by Mahbubul Haq and Amartya Sen, to which Jolly/Ray (2006) note that:

> The human security means safety for people from both violent and non-violent threats. It is a condition or state of being characterized by freedom from pervasive threats to people's rights, their safety, or even their lives. (p. 4)

Jolly/Ray (2006) identified seven core elements of human security – economic, food, health, environment, personal, community and political security – which reflect the basic needs of human life. Sundarbans contributes to several of these core elements of human security in the study area. The participants' note that the forest saves people from loss, ameliorates the threats of natural disasters and supports livelihoods with the security of economy, food, health and environment, as well as individual and collective sociopolitical life.

The forest resources form the most significant contribution of Sundarbans to human security, as the local population relies on the forest's water resources for their everyday life, food and shelter, whilst the income-producing activities – including fishing and collecting wood, honey and *pona* – have evolved over hundreds of years among the local population. Sundarbans also supplies the raw materials for other growing fields of employment, such as the shrimp culture projects in nearby districts. As such, Sundarbans has a pivotal role in protecting Shora's natural environment by absorbing carbon dioxide to ensure a reduction in temperature, most necessary in times of climate change. Although Sundarbans also protects the local region from losses arising from natural disasters like cyclones, its locality near the coastal regions renders it susceptible to frequent disasters emerging from the Bay of Bengal. Although the area suffered immeasurable losses, Sundarbans absorbed much of the impact of Sidr and Aila.

One focus group participant stated that during the cyclones, the people working around the forest took shelter in the forest, which acts like the calm within a cyclone centre. Not only does this protect the locality from immediate loss during disasters but during post-disaster recovery, it also provides support and resources, including *golpata*, bamboo, *shan*, wood, *bet* and a source of income. However, the participants also described the negative aspects, including threats to life and property, and limiting people to risky occupations. Although the forest provides livelihoods for the local population, tigers, snakes and crocodiles living in the forest and river *threaten the lives* of the forest users. The majority of women who use the forest lose their husbands to tiger attacks. Post Sidr and Aila, the forest ecology has broken down and, from observations, it is apparent that the forested portion surrounded by the river bank is mostly barren of trees and animals. *Threats to property* include the losses of property from the disasters and local wild animals. After the cyclones, the local population faced a food shortage and the wild animals could not satisfy their hunger from food within the forest, consequently moving out of the forest and into nearby villages in search of food. This threatened the lives of both the local population and their livestock, such as cattle.

Participants strongly believe that limiting people to *risky occupations* is a serious concern for their survival in Shora. The current infrastructure, facilities, social norms and values bind those involved in the traditional professions to their existing circumstances rather than allowing them to seek livelihoods elsewhere or in a safer manner. The local geography and infrastructure are not in favour of developing businesses, as the focus group participants reveal an intimate bond between the local population and the forest – 'the Sundarbans is our mother, our lives, our future and our love'. Another participant added, 'Suppose I am the head of my family, so the Sundarbans is in this zone.'

Since the 1950s, there have been unprecedented climatic changes, influenced by human beings, threatening the future world with natural resources scarcity and pollution (Intergovernmental Panel on Climate Change National Research Council 2010, cited in Tietenberg/Lewis 2016). As human-induced changes affect both humans and ecosystems, this is a significant issue in environmental discourse (Rogers et al. 2007). In the summer of 2003, increased heatwaves caused thousands of deaths in Europe, providing evidence of the changing climatic situation. The combination of smog and warm weather can negatively affect human lives, and rising temperatures will result in mass migration and loss of life, particularly in developing nations, hence creating an uneven loss of life.

The countries with limited greenhouse gas production but less adaptation capacity will be the worst affected by climate change, as will countries whose rising demands outstrip resource supply. Around 50% of Africans and 75% of Asians lack an adequate supply of safe water (UN, cited in Tietenberg/Lewis 2016), whilst low-lying countries are susceptible to rising sea levels. Tietenberg/Lewis (2016) note local and international policies should focus on the changing situation of the climate, and this focus should integrate food, water and energy security concerns in research and development (Millennium Project 2005; Loring et al. 2013). Local-level dynamics of food, water and energy are more complex than their general theories and studies, hence Zolotukhin et al. (2017) suggest that individual or state initiatives to confront climatic change are insufficient and that unified steps with global perspectives are necessary. Although there are geographical and administrative boundaries, large-scale environmental policy is required at local and global levels to contribute to the stabilization of administrative formulas (Zolotukhin et al. 2017).

To understand the affiliation between the environmental component and comprehensive security, Westing (1989) categorizes *Non-extractive Resources* – land, soil, water and atmosphere – and *Extractive Resources* – non-renewable and renewable resources. This is similar to the participants in the focus groups identifying three categories of environmental resources. In the *forest and wildlife* category, the forest and its inhabitants include trees, birds and animals such as deer, tigers, snakes and monkeys. The *rivers and water animals* category includes rivers and canals, fish, snails, crocodiles and water snakes, whilst the third category refers to the people or dependents related to the *process of human security*.

Participant: 'Shora's ecological setting has turned into an unstable one since it historically faced cyclones in the years 1975, 1988, 1991, 2005 and 2009, resulting in the diminishing of the inhabitations, including the livelihood of majority of the people.' (A female participant, age 45.)

Participant: 'Without this forest, we couldn't live here; it's the forest that cares much for our survival in every regard.' (A male participant, age 59.)

In addition, the participants strongly believe that Sundarbans contributes to protecting them from the environment and seasonal weather, including extreme cold in the winter and absorbing the increasing summer heat.

Participant: 'The forest is our natural oxygen factory. During wintry months, we tolerate bitter cold as we cannot afford to purchase expensive warm clothes, but generally we use ripper and blanket to warm ourselves. It is the forest which reduces the level of cold and keeps the area a bit warmer. In summer, on the other hand, we breathe pure oxygen and it keeps us fresh and beautiful.' (A male participant, age 55.)

Participant: 'The mangrove tree of the Sundarbans keeps the environment fresh and alive, and it makes our body beautiful.' (A female participant, age 65.)

Both women and men choose Sundarbans as their source of income and livelihood, and thus they interact with the forest through building kinship bonds.

Participant: 'Both men and women go to the Sundarbans for their living. In particular, the women who have lost their husbands in the forest have no other choice but to go to the forest or fishing. That's why the friendship between the forest and men and the nearby inhabitants deepens every day.' (A male participant, age 72.)

Whilst asking questions to document participants' views on the volatile local ecological setting, they distinguish between pre-Sidr and -Aila and post-Sidr and -Aila. All of the participants described Aila as cruelly affecting both coastal villages and Sundarbans, whilst Sidr caused damage to an extensive number of forest trees. After Aila, the floodwaters lasted for about three years at Shora, hampering the growth of corn and displacing the local population, who had no other choice but to live on the nearby embankment. Many relocated to the district town for residential support and in search of better income opportunities.

A male participant: 'On the day of Aila, it started to rain followed by a violent storm. The velocity of wind increased over time and destroyed the dam. As a result, water entered the villages, farms and fields, and washed away the people, houses and belongings. That was a nightmare and I don't want to remember it again.'

A female participant: 'Aila caused a great loss to our lives; as a result, we are afraid to get back to our natural life and we will have to survive with others' help.'

When asked about their views on restoring the Sundarbans ecology, one of the informants emphasized that the natural process of restoration would have happened, even without any deliberate human intervention.

Participant: 'Even after the Sundarbans were affected by Aila and Sidr, the trees would restore the vigour of life all by themselves as they can grow in salinity. But here at Shora the trees can't survive in salinity, so the area looks like a desert with no crops and trees left.'

In comparing the Shora's ecological setting in two situations, an informant was vocal in describing her experience:

Participant: 'In the years before the onslaught of Aila and Sidr, there had been trees everywhere in the locality. But now most of them have vanished. It was a well-ventilated atmosphere and a sound environment, but now it has been devastated with the environment being contaminated. The weather has turned hot, and there is a spate of skin diseases. After Sidr and Aila the environment and the weather and climate have changed significantly. During monsoon there is less rain, and in summer the scorching sun makes our life difficult. The region has been facing an increase in waterborne diseases like diarrhoea, lesions and the like. If there were enough trees like before, I think we would not have to suffer as much as we do now.'

According to the informants, in both cases, the forest was devastated with great losses to the trees, birds and animals, impacting the whole ecology, but during Sidr, the forest was more affected than the locality. Trees were stripped of their leaves, fish died in the river and wild animals died in the forest. All of the participants believed that, prior to the cyclones, each and every property in the area was in better condition than in the current state. The local population had lost their households, cattle and plantations, including coconuts, mangos, *shofeda*, dates, *kul*, *jaam*, *litchis*, jackfruits and most importantly the happiness in everyday life. Prior to the disasters, they used to cultivate vegetables like *puishak, dengashak, demushak, rangashak*, eggplants, potatoes, *folkopi, patakopi, raddises*, ladies' fingers and *kachurmukhi* in the fields, but now almost nothing grows. In addition, they produced paddy, namely *potinai, goti, goran, ghunshi, chaprali and hira* in their lands, but due to the loss of the fertility of the fields, now no crops are produced. The post-cyclonelandscape has left a devastated wasteland, with few trees left in the locality, fields with no crops and dry paddies. This has increased salinity, creating the worst life situation in decades. One of the participants shared her view that in any other disasters, it was easy to cope or adapt to the new situation because there were crops in the fields. The poor would get a job working in the fields of wealthy people. However, as the fields are now infertile, there are no work opportunities for the poor.

5.3 Environmental Security at Shora

Environmental security as a discourse links natural and anthropogenic processes (Richmond 2018). According to Krakowka et al. (2012), environmental security is comprised of environmental processes that, first, promote instability and undermine governments, and secondly push civil conflict. Richmond (2018) found that in sub-Saharan Africa, there are predominantly three components in an urban environment that influence environmental security. First, poor living conditions in regional and rural areas may push people to move to urban areas in the quest for better living conditions. Second, population growth inside the city has negatively affected the environmental resources. Lastly, poor people become marginalized and vulnerable to natural hazards. Richmond (2018) holds that as these components create additional problems for administrative and other agencies, there is no alternative other than rethinking and reconsidering the existing urban policies.

Food security is a long-term challenge for humans, and global food insecurity is increasing (Acevedo et al. 2018), hence environmental preservation, along with agricultural productivity, is a key concern. The Food and Agriculture Organization (FAO 2006) proposed four dimensions for food security: availability, access, utilization and stability. According to Bruinsma (2009), there might be an increase in agricultural production by 2050, but this may be tempered by the diverse reasons for food insecurity, including poverty and lack of accessibility (Scanlan 2001), whilst increased production may be affected by aquifer depletion, climate change, soil erosion and falling water tables (Acevedo 2011). To reduce these problems, Acevedo et al. (2018) believe the capacity of food supply should be increased with adaptations and innovations to economic planning, and by integrating science and technology through collaborative approaches among natural sciences, social sciences and engineering, as there is a key connection between food security and environmental security.

According to the Millennium Project (2009) report on 'State of the Future', environmental security is the relative safety from environmental dangers. These dangers are caused by natural events or human processes, such as ignorance, accident, mismanagement or design, and they originate within or across national borders (Glenn et al. 2009). The Millennium Project defines environmental security as environmental viability for life support and features three elements: (I) preventing or repairing damage to the environment, (II) protecting the environment due to its inherent moral value, and (III) preventing or responding to environmentally caused conflicts. However, the focus of the study has been on the first two issues, preventing or repairing damage and protecting environment due to its inherent moral values. The necessity of environmental security has been a crucial issue for the present condition of the study region.

First, frequent natural disasters like Sidr and Aila devastate the lives of the local population. As the forest is their immediate source of income, it operates as a support, and as the demands increase after a disaster, resource exploitation begins. Second, disasters like Sidr and Aila have made the forest too vulnerable for its survival and the situation demands care for the forest itself, not just support for the local population. Moreover, the animals of the forest are insecure, due to the greed of the forest users. Sometimes people hunt exotic animals and collect resources in an aggressive, unsustainable manner. The unstable environment in both the locality and the forest increases demand for support through resources and care, and increases the rate of resource extraction. Moreover, climatologists predict that the salinity intrusion in the local land makes environmental security an urgent issue, and thus increases the need to take immediate action for the betterment of the existing ecological setting in the coastal areas. Therefore, the necessity of ensuring environmental security is obvious in every measure.

Towards the end of the focus groups, the participants became emotional and optimistically reported that they intended to lead lives free of risks and environmental risks at Shora. A healthy forest ensured livelihood support before Sidr and Aila, but now the alternative income opportunities to reduce forest use have diminished, as have the payments for collected resources. The participants emphasized that the corrupt political leaders of Shora only visit them when they require votes, and local political leaders attempted to intervene when media outlets visited the area to hear their messages, likely restricting dissemination to government agencies. As such, the voice of the poor informants' lives is systemically ignored.

Participant: 'At present, nobody listens to our voice, and almost no one cares to take steps for our safety. We voted for the leader, hoping for an overall improvement in our lives, including an infrastructural change in the village so that each inhabitant can benefit from communication facilities. But the leader has broken the commitment and now rarely comes to pay any heed to the problems of our risky life. We need not brick-built houses but fertile land and fertilizer for food security, proper medical support during outbreaks of disease, easy access to the disaster-connected information, and protection of lives from the cyclones and natural calamities.' (A female participant, age 35.)

Although Bangladesh emits very low levels of greenhouse gas emissions, it is one of the countries most affected by climate change (Kaisa/Islam 2018). As Bangladesh is one of the leading countries of aquaculture production, with shrimp being a large export market, shrimp farming is the main source of income for the people of the coastal region. However, climate change is hindering shrimp farming because cyclones, storm surges, increased temperatures, drought, heavy rainfall, salinity, riverbank erosion, a rise in sea level and other natural disasters frequently attack the coastal areas of Bangladesh where shrimp is cultured. Consequently, the aquaculture production industry and its farmers and traders are some of the most affected victims (Kaisa/Islam 2018).

As shrimp aquaculture is economically viable for middle and higher class people, agricultural land is being purchased from poorer people, and this expansion of shrimp farming is rendering already marginalized people socially and economically vulnerable (Abdullah et al. 2017). Additionally, Afroz et al. (2017) maintain that small-scale farmers were excluded from their land by large-scale shrimp farmers. However, shrimp farming in coastal Bangladesh has a devastating effect on Sundarbans mangrove forests (Ahmed et al. 2018). To avoid this problem, an organic shrimp farming strategy has been proposed (Ahmed et al. 2018).

5.4   Women's Role in Protecting the Resources of the Sundarbans Forest

In 'Women and the Environment', the UNEP (2005) quoted a peasant woman from the Gaibandha region of Bangladesh – 'Life is a whole, it is a circle. That which destroys the circle should be stopped. That which maintains the circle should be strengthened and nurtured.' FAO (2014) notes that the gender division in forestry due to various social, cultural, economic and institutional reasons results in women receiving less access to and control over forest resources. This is despite women having specialized knowledge about the use and usefulness of forest resources and their contribution to their household's food security (FAO 2014). Although the inclusion of women in the decision-making process of the forest sector yields significant outcomes in forest resource management (FAO 2014), women are rarely considered important in policymaking, compounding the gender discrimination as their lives mirror the circle of the life of the environment. As the caregiving role of women is interrelated with the role of the environment, it should be kept alive and nurtured, for the healthy maintenance of the circle of the life and the environment (Glenton et al. 2010). The studies conducted by Shambel in 2010 in the Philippines and in 2012 in Ethiopia reveal that women who use the forest develop their own informal approaches to solving natural resource extraction impediments. They have developed an informal institutional framework using their traditional knowledge of the sustainable use of natural resources. The women in Sorsogon, Philippines believe that, in order to protect themselves, their families and the next generations, it is crucial to protect the environment around them (Guiriba 2010). The outcomes of this study from the Philippines and Ethiopia indicate that, despite their gender roles, where women who use the forest are found to be more careful in their actions whilst collecting resources from the forest, they play pivotal roles in facilitating forest resource sustainability. The women use their traditional knowledge to develop the institutional framework and management system over the exploited zones, and do so through both individual contributions and collective actions.

Although the women who use the forest are treated according to the local Islamic religious leader (e.g. Huzur) in the region, the reality is they are fighting for survival. hence they are forced to overcome social barriers just to go to the forest. As the patriarchal system establishes the gendered division of labour, men obtain more benefits from the forest resources, hence women are under-represented and exploited in livelihood activities. The participants believed that women have a natural instinct to take care of their surroundings, and thus minimize the negative impacts of their activities whilst extracting resources from the river and forest. A female participant noted that when she sees a tree about to fall into the river, she puts it in a place where it might survive.

Whilst women and men have caregiving notions to act in accordance with the environment, it varies in different situations. There is a normal tendency among the people, regardless of gender, to leave live fish in the river after collecting the necessary *pona* by *ochol* so that these fish can grow further. However, there are practices that show women care more than men, when they take care in cutting down trees from the forest by first searching for dead trees or branches from the ground for their fuel woods. In contrast, men care more than women about the way trees are cut; and men follow a certain methodology when cutting the branches of a tree, which allows the trees to grow further in the future. Comprising the responses and data from the field, it is apparent that women take better care of the local ecosystem. They go to the Sundarbans in groups to minimize the impacts of over-extraction of forest resources and initiate tree planting, hence due to their natural instincts, women's role in extracting resources shows a sustainable approach.

5.5 Conclusion

The findings from the gathered data mostly unfold the sources of economic security, as well as people's perceptions of environmental security. With both the positive and negative aspects of human security, Westing's categorization of environmental resources in relation to comprehensive environmental security demonstrates an intimate bond between the local population and the forest. This is revealed through the participants claiming that the impacts of Sidr and Aila have affected the human–environment interactions in the locality. Therefore, the local population realizes the core role the forest holds in environmental security stability in the study location. This study reveals that women take special care of the forest, as they would care for their families, even though sociocultural and religious practices in the location cause women to downplay their roles in the forest.

References

Abdullah, A.N.; Myers, B.; Stacey, N.; Zander, K.K.; Garnett, S.T. (2017), 'The Impact of the Expansion of Shrimp Aquaculture on Livelihoods in Coastal Bangladesh', *Environment, Development and Sustainability*, Vol. 19, pp. 2,093–2,114. https://doi.org/10.1007/s10668-016-9824-5.

Acevedo, M.F. (2011), 'Interdisciplinary Progress in Food Production, Food Security and Environment Research', *Environment Conservation*, Vol. 38, No. 2, pp. 151–171.

Acevedo, M.F.; Harvey, D.R.; Palis, F.G. (2018), 'Food Security and the Environment: Interdisciplinary Research to Increase Productivity While Exercising Environmental Conservation', *Global Food Security*, Vol. 16, pp. 127–132.

Afroz, S.; Cramb, R.; Grunbuhel, C. (2017), *Exclusion and Counter-Exclusion: The Struggle over Shrimp Farming in a Coastal Village in Bangladesh*, The Hague: Institute of Social Studies.

Ahmed, N.; Thompson, S.; Glaser, M. (2018), 'Integrated Mangrove-Shrimp Cultivation: Potential for Blue Carbon Sequestration', *Ambio*, Vol. 47, pp. 441–452. https://doi.org/10.1007/s13280-017-0946-2.

Bruinsma, J. (2009), 'The Resource Outlook to 2050. By How Much Do Land, Water Use and Crop Yields Need to Increase by 2050? Session 2: The Resource Base to 2050: Will There Be Enough Land, Water and Genetic Potential to Meet Future Food and Biofuel Demands?', in: FAO (ed.), *Expert Meeting on How to Feed the World in 2050*, June 2009, Rome: UN FAO.

FAO (2006), *Food Security, Policy Brief*, June 2006, Issue 2. http://www.fao.org/forestry/13128-0e6f36f27e0091055bec28ebe830f46b3.pdf (April 2017).

FAO (2014), *Women in Forestry: Challenges and Opportunities*, Rome: United Nations.

Glenn, J.C.; Gordon, T.J., Florescu, E.; The Millennium Project Team (2009), *The Millennium Project Report on State of the Future*, Washington, DC: United Nations.

Glenton, C.; Scheel, I.B.; Pradhan, S.; Lewin, S.; Hodgins, S.; Shrestha, V. (2010), 'The Female Community Health Volunteer Programme in Nepal: Decision Makers' Perceptions of Volunteerism, Payment and Other Incentives', *Social Science & Medicine*, Vol. 70, No. 12, pp. 1,920–1,927. https://doi.org/10.1016/j.socscimed.2010.02.034.

Guiriba, G.O. (2010), 'The Role of Women in Environmental Conservation in Sorsogon Province, Philippines', in:*4th Asian Rural Sociology Association (ARSA) International Conference*, Philippines, pp. 106–112.

James, P. (2014), 'Human Security as a Left-over of Military Security, or as Integral to the Human Condition', in: Paul Bacon and Christopher Hobson (eds.), *Human Security and Japan's Triple Disaster*, London: Routledge, p. 73.

Jolly, R. and Ray, B.D. (2006), *The Human Security Framework and National Human Development Reports: A Review of Experiences and Current Debates*, NHDR Occasional Paper 5, United Nations Development Programme (UNDP), Brighton: Institute of Development Studies.

Kaisa, S.M. and Islam, M.S. (2018), 'Impacts of and Resilience to Climate Change at the Bottom of the Shrimp Commodity Chain in Bangladesh: A Preliminary Investigation', *Aquaculture*, Vol. 493, pp. 406–415.

Krakowka, A.R.; Heimel, N.; Galgano, F.A. (2012), 'Modeling Environmental Security in Sub-Saharan Africa', *Geography Bulletin*, Vol. 53, No. 1, pp. 21–38.

Loring, P.A.; Gerlach, S.C.; Huntington, H.P. (2013), 'The New Environmental Security: Linking Food, Water, and Energy for Integrative and Diagnostic Social-Ecological Research', *Journal of Agriculture, Food Systems, and Community Development*, Vol. 3, No. 4, pp. 55–61. http://dx.doi.org/10.5304/jafscd.2013.034.005.

Millennium Project (2005), *Investing in Development: A Practical Plan to Achieve the Millennium Development Goals*, New York: United Nations. Viewed 17 April 2019 http://siteresources.worldbank.org/INTTSR/Resources/MainReportComplete-lowres%5B1%5D.pdf.

Paris, R. (2001), 'Human Security – Paradigm Shift or Hot Air?', *International Security*, Vol. 26, No. 2, pp. 87–102.

Richmond, A.K. (2018), 'Water, Land, and Governance: Environmental Security in Dense Urban Areas in Sub-Saharan Africa', in: F. Galgano (ed.), *The Environment-Conflict Nexus: Climate Change and the Emergent National Security Landscape*, Basingstoke: Springer Nature. https://doi.org/10.1007/978-3-319-90975-2_6

Rogers, P.; Jalal, K.F.; Boyd, J.A. (2007), *An Introduction to Sustainable Development*, London: Routledge.

Scanlan, S.J. (2001), 'Food Availability and Access in Lesser-Industrialized Societies: A Test and Interpretation of Neo-Malthusian and Techno Ecological Theories', *Sociological Forum*, Vol. 16, No. 2, pp. 231–262.

Tadjbakhsh, S. and Chenoy, A.M. (2006), *Human Security: Concepts and Implications*, London: Routledge.

Tietenberg, T. and Lewis, L. (2016), *Environmental & Natural Resource Economics*, 10th Edn., New York: Routledge.

United Nations Environment Programme (2005), *Annual Evaluation Report 2004*, Evaluation and Oversight Unit.

Westing, A.H. (1989), 'The Environmental Component of Comprehensive Security,' *Bulletin of Peace Proposals*, Vol. 20, No. 2, pp. 129–134.

Zolotukhin, V.M.; Gogolin, V.A.; Yazevich, M.Y.; Baumgarten, M.I.; Dyagileva, A.V. (2017), 'Environmental Management: The Ideology of Natural Resource Rational Use', *Earth and Environmental Science*, Vol. 50, 12–27. https://doi.org/10.1088/1755-1315/50/1/012027.

Chapter 6
Implications of the Gendered Knowledge About the Sundarbans Forest at Shora and Beyond

Abstract Shora as an island village is adjacent to the Bay of Bengal and Sundarbans mangrove forest. Lives and livelihoods of villagers of Shora have been completely dependent to the forest resources and rivers for centuries. Earning a livelihood in the Sundarbans is extremely risky for life. Fighting with a series of natural disasters, this chapter offers knowledge demonstrating how forest-going perceptions and behaviours of Shora people were impacted by cyclones Sidr and Aila. This chapter shows a cursory analysis of the implications of gendered knowledge about the Sundarbans Forest at Shora and beyond.

Keywords Gendered knowledge · Gendered politics · Sundarbans forest

This study documents the ecological knowledge of Shora women and men living in and around the south-west of the Sundarbans forest region by showing the ways women use the forest and its resources. Stevenson (1996) defines traditional ecological knowledge (TEK) as comprising three interrelated components: specific environmental knowledge, knowledge of ecosystem relationships, and a code of ethics. The TEK focuses on the environmental knowledge gained through people's experiences and traditions (Usher 2000). Whilst Huntington (1998) terms TEK as an experiential knowledge system of a rural community, Houde (2007), notes that there are six faces of traditional ecological knowledge – management systems, past and present land use, ethics and values, factual observation, cosmology, and culture and identity. This study has explored the ways in which the socio-cultural identity of Shora females and males influenced their use of Sundarbans forest resources prior to and after cyclones Sidr and Aila. In doing so, the study addressed the following questions:

- What are women's and men's perceptions about the Sundarbans, its resources and changes in the use of forest resources due to cyclones Sidr and Aila?
- How do women actually use forest resources?
- What are women's and men's perceptions of environmental security, as it relates to the forest and their region?

© The Author(s), under exclusive license to Springer Nature Singapore Pte Ltd. 2019   85
S. Roy, *Climate Change Impacts on Gender Relations in Bangladesh*,
SpringerBriefs in Environment, Security, Development and Peace 29,
https://doi.org/10.1007/978-981-13-6776-2_6

This research study has highlighted how the long-established faiths and beliefs around the Sundarbans forest, and the pattern of using forest resources, are transformed, and transferred from the Shora's elder inhabitants to the younger generations. Although the law enforcement officials for the forest are supposed to protect the forest from commercial usage, the local forest rangers, and the dishonest Shora men, use the Sundarbans forest for such purposes. The study shows that men, contrary to women, go to the forest at least twice a day when there is the flow of high water and low water filling the forest canals. Before Sidr and Aila, the long-established social norms and negative labelling towards women reinforced domesticity, but after Sidr and Aila, married women are gradually working outside the home in the local labour force for small-scale income generation activities at Shora. In contrast, the local mainstream community excludes divorced women, who have to withstand social criticism to access the nearest part of the forest.

The study reveals that forest-dependent women apply their own knowledge to catch a wider variety of fish species and the ways they use unsharpened instruments to remove the dead branches of mangrove trees. Their knowledge extends to how forested non-timber products are used the household diet, or sold in the market to earn a sustainable income.

Even though married women spend their time processing the forest resources gathered by their husbands, their role is not recognised because their invested physical labour and time occurs in the household domain, and is thus perceived as weaker and less important. This reinforces the men's position as breadwinner, with support from the traditional Muslim values in Shora. Consequently, these conditions deprive women of a share of the money and push them to the periphery of society. This oppressive social norm of disrespect and deprivation prevents both married and divorced women from assuming a socio-political identity, as they must negotiate hierarchal power in and outside the home, rather than confronting their oppression with their experiences and arguments. However, as women directly conserve the forest, this inspires the men, particularly their husbands, to patrol and protect the village from wild animal attacks. This environmental responsibility demonstrates Shora community members' sense of care towards the wild animals and resources in the Sundarbans forest. Despite the reinforced gender roles, in the aftermath of Sidr and Aila, the Sundarbans' ecology deteriorated, creating desert-like conditions. As the ecosystems of the region are in a transitional phase for natural recovery, Shora women and men are in need of a non-forest-dependent source of income to give the Sundarbans' ecosystem a chance to recover.

According to Harding's Standpoint theory, described in Chap. 3, the female participants occupy two kinds of standpoint: (a) risky situations during their stay in the forest, and (b) deprivation in the home by husbands and at market by fish dealers. These standpoints are influenced by women's vulnerability due to social forces and the cyclones, whilst the risky situations in the forests are primarily caused by pirates and wild animals. Despite the situations and standpoints, women have access to the forest, and it is there that a social stratification develops among women and men based on income earned from forest resources.

It has been discussed how and why women's voices are not heard or are kept concealed, thus the study reflects the way that marginalized women's knowledge of production from the grass-roots level, including their conservationist attitudes towards this forest, critically reflects the Feminist Political Ecology Theory. The study also highlights that men's toil during honey-collection and Golpata-cutting requires a stay of more than a month inside the Sundarbans, thus Shora males continue to struggle in the quest to secure a livelihood for their family members.

This study also shows an ambiguity, as it does not document the gender relations of the forest dependents, consider women's and men's invested times in collecting forest resources from the Sundarbans, its processing mechanism, or how the local historical politics of distribution of the gained financial benefits deprive women of their payment. The study could have explored the local population's gendered ideology, as Radel (2012) did in terms of struggle for the use of Sundarbans forest resources at the household and village levels or the way Sidr and Aila survivors were internally migrated from Shora to a nearby town for work for their survival in the region. Documenting the connectedness between an indigenous community such as the Munda community and the Sundarbans was beyond the scope of this particular study. Time and monetary constraints prevented the researcher detailing the participants' struggle for a safe and secure livelihood related to the Sundarbans, or unravelling the mess of daily politics that hinder women in claiming the rewards of the invested labour.

The researcher learned much about the ingenious perception of the Sundarbans, the diverse use of the forest, women's interactions with the forest, and peoples' construct of the role of the forest for environmental security. The researcher wishes to return to Shora for further research to address the aforementioned research gaps.

References

Davis, A. and Ruddle, K. (2010), 'Constructing Confidence: Rational Skepticism and Systematic Enquiry in Local Ecological Knowledge Research', *Ecological Applications*, Vol. 20, No. 3, pp. 880–894.

Dickinson, J.L.; Shirk, J.; Bonter, D.; Bonney, R.; Crain, R.L.; Martin, J.; Phillips, T.; Purcell, K. (2012), 'The Current State of Citizen Science as a Tool for Ecological Research and Public Engagement', *Frontiers in Ecology and the Environment*, Vol. 10, No. 6, pp. 291–297. https://doi.org/10.1890/110236.

Houde, N. (2007), 'The Six Faces of Traditional Ecological Knowledge: Challenges and Opportunities for Canadian Co-management Arrangements', *Ecology and Society*, Vol. 12, No. 2, p. 34. http://www.ecologyandsociety.org/vol12/iss2/art34/.

Huntington, H.P. (1998), 'Observations on the Utility of the Semi-directive Interview for Documenting Traditional Ecological Knowledge', Vol. 51, pp. 237–242.

Radel, C. (2012), 'Gendered Livelihoods and the Politics of Socio-environmental Identity: Women's Participation in Conservation Projects in Calakmul, Mexico', *Gender, Place & Culture: A Journal of Feminist Geography*, Vol. 19, No. 1, pp. 61–82.

Stevenson, M.G. (1996), 'Indigenous Knowledge in Environmental Assessment', Vol. 49, pp. 278–291.

Usher, P.J. (2000), 'Traditional Ecological Knowledge in Environmental Assessment and Management', *Arctic*, Vol. 53, pp. 183–193.

Glossary of Bengali Words

Aila A cyclone

Allah Muslim's God

Bada Local name of Sundarbans

Bagh Tiger

Bagdha Shrimp only found in saline water

Baishak First month of Bengali year

Bedhobapolle Shelter home for divorced women

Borse Fishing instrument

Chattro Last month of Bengali year

Chira Bangladeshi food

Chati Large shrimp

Dinghy Tiny boat

Dowa Muslim's prayer

Easy Bike Three-wheeled motor vehicle

Furi Local fish collector

Gita Holy volumes for Hindu people

Jal Net

Jala-Baoalie Female fishing community

© The Author(s), under exclusive license to Springer Nature Singapore Pte Ltd. 2019
S. Roy, *Climate Change Impacts on Gender Relations in Bangladesh*,
SpringerBriefs in Environment, Security, Development and Peace 29,
https://doi.org/10.1007/978-981-13-6776-2

Kawra Fruit found in Sundarbans

Kholpatua Name of a river

Kortabakte Male head

Lakre Fuel

Lungi Man's cloth

Mabonbibi Goddess

Mach Fish

Mal Local name of Sundarbans

Mawali Honey gatherer

Mayabiehorin Beautiful Deer

Meku Water-insect

Modantak Bangladeshi bird

Muri Bangladeshi food

Ochol Technique for separating fish from water-insect

Pantavat Fermented boiled rice

Passea Salty sea fish

Pona Shrimp

Powa Name of a fish

Renu Small shrimp

Sidr Name of a cyclone

Shora Name of a village

Sofeda Bangladeshi fruit

Taka Official currency of Bangladesh

Union The lowest level of administrative unit of Bangladesh

Appendix A
In-depth Interview Guide

This document is a guideline for conducting unstructured interviews during the fieldwork.

Sundarbans Forest

- Describe, in detail, your understanding of the Sundarbans.
- When and how did you become acquainted with the forest?
- Do you require any permission to access the forest? Share your personal experience.
- What is the preferred time for going to the forest?
- Who goes to the forest, and for what?
- Tell us about the clothes the women and men wear when they go to the forest.
- What are the rituals which the area's inhabitants practise before going to the jungle?
- Outline the available resources of the Sundarbans Forest, and what types of forest resources you collect.
- Tell me the name of the instruments you use when you collect resources from the forest.
- How do you shift the collected resources from the forest?
- According to you, who is involved in processing the collected resources for market, and how does the process function?
- What are the benefits people obtain from the forest?
- Describe the relationship the area's inhabitants have with the forest, and how it changes over time.

S. Roy, *Climate Change Impacts on Gender Relations in Bangladesh*,
SpringerBriefs in Environment, Security, Development and Peace 29,
https://doi.org/10.1007/978-981-13-6776-2

Informant's Behaviour with the Forest and its Resources

- Are there any differences between using instruments and collecting forest resources? If so, share your experiences.
- How do the women collect fish, leaves and timbers and how do they interact with the collected resources at home?
- To what extent, and by whom, is the forest used in a sustainable way?
- Tell me how the forest resources are sold in the market, and who benefits financially.
- Tell us about the social barriers women face when they go to forest, and when they attempt to sell the forest products in the local market.
- What are the obstacles the women face when they influence the familial decision-making process?
- Tell me in detail about the people's forest conservation activities.

Environmental Security

- Do you feel that you are living in a secure environment? How was the ecological system of the village when you were a younger? Describe the crops and habitations of the people of Shora.
- Describe the recent environmental degradation that has occurred in the area.
- According to you, what are the likely reasons for the degradation of the local ecosystem, as well as the forested ecology?
- Tell me your experiences about the cyclones that have occurred at Shora.
- Tell me in detail about the scarcity or availability of the environmental resources required for leading a safe life.
- Describe the relationships between the people and the forest.
- Share with me how the cyclones hampered the environmental security of the area.
- What is the relationship between the Sundarbans forest and environmental security?

Appendix B
Keywords Used in Focused Group Discussion

Sundarbans[1]
Religious practice
Men
Women
Widow
Divorced women
Bidobapolle
Condition of the forest
Corruption
Use of the forest resource
Livelihood
Forest-going habits of men and women
Allocation of the benefits gained from the forest resources
Forest conservation
Tiger attacks
Patrolling system
Sidr
Aila
Environmental security
Ecological protection
Forest and people's relation.

---

[1]For the gathering of rich data and convenience of the participants, I used keywords during focus group interviews.

© The Author(s), under exclusive license to Springer Nature Singapore Pte Ltd. 2019
S. Roy, *Climate Change Impacts on Gender Relations in Bangladesh*,
SpringerBriefs in Environment, Security, Development and Peace 29,
https://doi.org/10.1007/978-981-13-6776-2

About the Author

Sajal Roy has been lecturing (although he is currently on study leave) in the Department of Women and Gender Studies at the Begum Rokeya University, Rangpur (BRUR), Bangladesh since 2014. He is currently enrolled as a Ph.D. candidate at the Institute for Culture and Society (ICS), Western Sydney University, Australia. Before commencing his Ph.D., he completed a Master of Philosophy (M.Phil.) in Gender and Development from the University of Bergen and a Master of Social Sciences (MSS) in Women and Gender Studies from the University of Dhaka. His doctoral thesis examines the ways that gender, marital status, religion and mobility intersect with the forest-based livelihood transformation of two contrasting coastal communities (Muslim and Munda Indigenous Communities) in south-west Bangladesh since Cyclone Aila in 2009. His documentary, entitled *Livelihood Diversity in the Sundarbans Forest*, has recently been screened at Lund University's Development Studies conference. Sajal's teaching and research interests include applied social research methods, community studies, gender and post-disaster recovery, feminist political ecology, and sustainable development goals. Sajal was awarded the prestigious *Meltzer Research Grant* from the University of Bergen in 2012 and a small-scale research grant (2015), jointly awarded by the Begum Rokeya University, Rangpur and the University Grants Commission of Bangladesh on feminist political ecology of gendered relations of the Shora forest community in the Bangladesh Sundarbans forest. He was a visiting researcher at the Nordic Institute for Asian Studies (NIAS), University of Copenhagen and International Centre for Climate Change and Development (ICCCAD), Independent University, Bangladesh. He also served as a Senior Research Associate at the Research and Evaluation Department of BRAC International in South Sudan, Sierra Leone and Liberia. As a young social worker and resource person, he has contributed for more than 3 years to a local NGO called *Coastal Development Organization for Women*, located in the district of Satkhira, Bangladesh.

Peer-reviewed Publications:

- Roy, S. (2018), 'Livelihood resilience of the indigenous Munda community in the Bangladesh Sundarbans forest' in: W. Leal and D. Ayal (eds.), *Handbook of Climate Change Resilience*, Springer Nature, pp. 1–22.
- Roy, S. (2018), 'Book review: Women of the storm: civic activism after Hurricane Katrina", *Journal of Gender Studies*, https://doi.org/10.1080/09589236.2018.1475322.
- Roy, S. (2017), 'Book review: Women and disasters in South Asia: survival, security and development', *Gender, Place and Culture*, pp. 1–2, Available: http://dx.doi.org/10.1080/0966369X.2017.1396679.
- Roy, S. (2018), 'Book review: *Living in the Anthropocene: earth in the age of humans'*, *Environment and History*.
- Roy, S. (2017), 'Book review: *Gendered Lives, Livelihood and Transformation: The Bangladesh Context'*, *LSE Review of Books*.
- Roy, S. (2018), 'Book review: Women and Girls: Vulnerable or Resilient?', *Journal of the Asiatic Society of Bangladesh (Humanities)*.

Address Sajal Roy, Institute for Culture and Society, Western Sydney University, Building EM, Parramatta South Campus, Locked Bag 1797, Penrith NSW 2751, Australia.
Email roysajal.wgs@gmail.com
Websites https://sites.google.com/site/sajalroybdasia/home and http://www.afes-press-books.de/html/SpringerBriefs_ESDP_29.htm

About the Book

A truly comprehensive introduction to the topic, *Climate Change Impacts on Gender Relations in Bangladesh: Socio-environmental Struggle of the Shora Forest Community in the Sundarbans Mangrove Forest* is an essential text for undergraduate and postgraduate students but also anyone wanting to better understand the complexities between gender, disasters and development in the coastal regions of Bangladesh. Sajal Roy's own deep knowledge, passion for the subject and accessible writing style provide great insights into the different cultural, social, political and environmental perspectives of this agenda. The combination of accessible introductory description along with the critical commentaries is difficult to pull off, but this book manages it with style.

> *Saleemul Huq*, Ph.D., Director, International Centre for Climate Change and Development (ICCCAD), Bangladesh, and Senior Fellow, International nstitute for Environment and Development (IIED), UK.

Elucidating the socio-environmental struggles by the Shora forest communities living in the Sunderbans areas, Sajal's book contributes immensely to the pertinent yet somewhat under-researched area of climate change and its nuanced impacts on gender relations in Bangladesh. Findings reiterate that empowering local communities leads to building communities' social capital, which they already possess, and possibly facilitates identifying community-led, gender-sensitive mechanisms in saving the forest. The book recommends concerted efforts by agencies and stakeholders in supporting longer-term initiatives such that climate change impacts on gender roles and relations can be positively realized, especially among communities like the Shora, who are exposed to multiple layers of vulnerabilities: climatic hazards, patriarchal societal construct, and inadequacies of resources.

> *Golam M. Mathbor*, Ph.D., Professor, School of Social Work, Monmouth University, New Jersey, USA.

Most climate adaptation strategies focus on sustaining ecosystems and meeting basic human needs. Invariably this involves concepts of past and future equity in terms of access and use of resources. Sajal Roy analyses the changes in gender relationships in the context of impacts on vulnerable communities and ecosystems in the Bangladesh Sundarbans. This

S. Roy, *Climate Change Impacts on Gender Relations in Bangladesh*,
SpringerBriefs in Environment, Security, Development and Peace 29,
https://doi.org/10.1007/978-981-13-6776-2

excellent research provides an essential dimension for understanding and directing local climate adaptation as well as broader strategies for climate justice.

> *Donna Craig*, Ph.D., Professor in Environmental Law, School of Law, Western Sydney University, Sydney, Australia.

Sajal's rich ethnographic analysis of the Shora Sundarban forest community provides valuable insights into the gendered dimensions of disaster resilience and livelihood security. In a time of anthropogenic climate change, such intersectional thinking is essential to future poverty-alleviation efforts. Scholars and practitioners of gender and development, rural livelihoods, climate change and development, South Asian development, human security, disaster resilience, forest-dependent communities, and feminist political ecologies will all find this book useful.

> *Dr. Kearrin Sim*, Lecturer and Program Convenor in Development Studies, College of Science and Engineering, Division of Tropical Environments and Societies, James Cook University, Australia.

This book is a useful contribution to the much-needed study space for understanding the gendered imbalances in the context of traditional knowledge, contribution and deprivation in sustainable use of resources in Sundarbans. The author explores vividly the causal relationship between the complex ecological processes and the socio-environmental struggles, which will be a valuable contribution to anyone interested in women's role in creating a sustainable future.

> *Dr. Namrata Bhattacharya-Mis*, Lecturer in Human Geography and International Development, University of Chester, UK.

The book by Sajal Roy on Sundarbans uncovers several myths about this largest mangrove forest in the world – the home of Bengal Tigers. The struggles for livelihood by the local people, especially Muslim women, who risk their lives while roaming in this forest are depicted in the book in a fascinating manner, which will be liked by any reader.

> *Professor Ataur Rahman*, Western Sydney University, Australia.

Index